Transformation of Survey of India

By

Prithvish Nag

Transformation of Survey of India

By Prithvish Nag

This book first published 2023

Ethics International Press Ltd, UK

British Library Cataloguing in Publication Data

A catalogue record for this book is available from the British Library

Print Book ISBN: 978-1-80441-238-1

eBook ISBN: 978-1-80441-239-8

Dedication

This book is dedicated to my parents,
Dr Pradyumna Chandra Nag and **Mrs Sucharoo Nag**

The author, taking over the charge as Surveyor General of India on 3rd December 2001 from Lt. Gen A.K. Ahuja

"The relevance comes from the fact that the scientific basis of this entire experiment has not changed while technology has changed. You see this in several areas. Our entire communication system today is electro-magnetic and digital, but the technology has changed. The digital revolution happened 20 years ago and is getting upgraded. The basic principles of mapping and surveying have set at that time and new technologies are coming into the picture. It is important because there is greater realisation that the domain in which maps will be useful has considerably broadened. Therefore we have to tell our people that we have a tradition in mapping so that everybody becomes familiar with the kind of knowledge maps provide them."

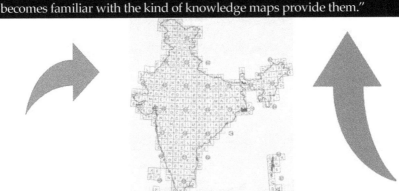

"Many countries have mapping organisations within which the production processes for topographical mapping date from colonial times. Such organisations have a hard time to keep pace, in terms of production rates, with the fast development of their country. Traditional approaches are too slow. I think that along with new mapping techniques we should develop new concepts for core data production and provision which avoid the slow processes of traditional map production. If we do not dwell too much on interpretation but instead correlate data geographically sufficiently correctly with GPS, then we can work on the provision of a good core dataset capable of being linked to other types of information, like vegetation and soil data. In training and conceptual development there should come a move away from traditional map concept still so often found in the definition of products."

Martien Molennar

Contents

About the Author

Dr. Prithvish Nag, M.Sc. (Gold Medallist), Ph.D. has been the Vice-Chancellor of the Mahatma Gandhi Kashi Vidyapith, Varanasi; Uttar Pradesh Rajarshi Tandon Open University, Prayagraj; Samoornanand Sanskrit Visvavidyalaya, Varanasi; and Deen Dayal Upadhyay Gorakhpur University. He was elected as a Fellow of the Royal Geographial Society, London and receipient of the Commonwealth Geographical Bursary on the 150 anniversary of the Royal Geographical Society.

Dr Nag was Senior Visiting Member of the School of Oriental and African Studies, University of London. He has been Scientist/Engineer in the Indian Space Research Organisation (ISRO). He was also Visiting Professor in the Department of Civil Engineering of the Indian Institute of Technology, Banaras Hindu University. He is an author and editor of 77 books and 135 research papers. His well known books are *Geography of India, Population Geography* and *Digital Remote Sensing*.

Dr Nag has been the President of the of Indian National Cartographic Association, Indian Association of Special Libraries and Information Centre, Institute of Indian Geographers, Indian Institute of Geomorphologist, Professor S.P.Chatterjee Memorial Foundation and Regional Science Association of India. At the International level, he was the Chairman of the International Cartographic Association (ICA), Commission on Population Cartography, UN / PCGIAP Committee on Institutional Strengthening for Asia and Pacific, International Steering Committee on Global Mapping (ISCGM) Working Group on Data Standardization; Full member of the International Geographical Union (IGU) Commission on Population Geography, and Member, Editorial Board, International Journal on Geographical Information System.

Dr. Nag was the Director in the National Atlas & Thematic Mapping Organisation (NATMO) of the Ministry of Science & Technology, Government of India, UN Consultant in Cartography in the Sultanate of Oman and the Surveyor General of India. He has been chairmen of several Department of Science & Technology (DST) Expert Committees, such National Geospatial Programme (earlier Natural Resource Data Management System), National Spatial Data Infrastructure, Geospatial Chair Professors; and Member of the DST/FIST Expert Committee on Earth & Atmospheric Sciences. He is the former Chairman, GIS Executive Advisory Committee of the Kolkata Municipal Corporation.

He was Principal Expert in the Nanjing Normal University, Nanjing Project on "Dynamic Mapping and Service for Regional Ecological Civilization based on Big Data", December 2019; and was honoured as an Expert in the Home of Zhejiang Academicians in Moganshan, P. R. China, 2019. At present, Dr Nag is the Director of the SHEPA Group of Educational Institutions in Varanasi.

सत्यमेव जयते
डॉ. एस. चंद्रशेखर
Dr. S. Chandrasekhar

ONE EARTH · ONE FAMILY · ONE FUTURE

सचिव
भारत सरकार
विज्ञान एवं प्रौद्योगिकी मंत्रालय
विज्ञान एवं प्रौद्योगिकी विभाग
Secretary
Government Of India
Ministry of Science and Technology
Department of Science and Technology

28th March, 2023

FOREWORD

I am pleased to learn that one of our colleagues has written a memoir of his tenure as Surveyor General of India. Dr Prithvish Nag has been associated with Department of Science & Technology, Government of India in different capacities. His experience would be valuable in knowing how an old scientific organisation can be transformed in to a vibrant one by meeting the expectations and challenges of the society and the country.

As the National Cartograhic organization, Survey of India, is spread all over country having a unique responsibility to develop geospatial data. Modern technology has helped in changing from printed maps to digital data having options for choosing different platforms for access and applications. Considering the new geospatial guidelines and policy, this organisation is to refocus its priorities. It has been entrusted to pilot the implementation of this policy under the aegis of the Department of Science & Technology. The government expects its greater role in the success of national mega projects like smart cities, digital India and also to be a partner in achieving the India's target of US 5-trillion-dollar economy. Strength from history is of course an asset which must be understood in a right perspective. Hence, the role of a book like this becomes important.

Survey of India is likely to produce a unique transformation model. World would like to see and watch how nearly a 250-year-old scientific organisation changes itself. It will prove to be a model for similar survey institutions elsewhere.

I am sure that this book will be appreciated by persons who are interested in transformation management and scenario building of scientific institutions.

(S. Chandrasekhar)

Preface

I was highly impressed by the documentations about different activities and procedures of Survey of India. These printed documents sometimes are known as records, manuals, papers, or memoirs. Mostly they were for internal use in order to have parity while mapping the massive mapping activities covering the undivided South Asia. This region includes India, Pakistan and Bangladesh and beyond including Afghanistan, Nepal, Bhutan Sri Lanka and Myanmar. Nevertheless, Survey of India was sometimes responsible to map areas like Hong Kong, Karakoram, northern Africa and the like. The record of mapping these areas is well preserved in different establishments of Survey of India, particularly in Dehra Dun and Kolkata.

The reports and documents, mostly printed, cover wide range of topics. They include annual reports of topographical surveys, account and results of operations of great trigonometrical survey (standard, major & the baseline), triangulation of Hong Kong and New Territories, selection records of trigonometrical and topographical surveys, tidal records, auxiliary tables for graticules of maps, report of land revenue administration of North Western Frontier Province (settlement), settlement report of Bannu district (now in Pakistan), sketches on services during the Indian frontier campaigns, and the like. The standard operating procedures of ancillary activities were also well documented. Some of them were concerning (a) rules of printing and binding, (b) norms for ministerial and lower sub-ordinate establishments, (c) fieldwork operations, (d) procedures of raising camps, and on similar issues.

Regarding the technical work, the *Handbook of Topography*, having different chapters in printed form were most important. The vital chapters were on *Introductory* (Ch I), *Constitution and Duties of a Survey Party* (Ch II), *Triangulation and its Computation* (Ch III), *Theodolite Traversing* (Ch IV), Plane-Tabling (Ch V), Fair Drawing (Ch VI), Trans Frontier Reconnaissance (Ch VII), Survey in War (Ch VIII), *Geographical Maps* (Ch XI), *Map Reproduction* (Ch X), *Forest Survey & Maps* (Ch IX), and *Air Surveys* (Ch XII). These chapters have written before independence and revised several times after independence as well. For example, the Chapter X on Map Production was first published in 1911 and the second edition in 1919. Further, the chapters on Triangulation and its Computation and Plane-Tabling were six times revised under the orders of the then Surveyor General of India, Brigadier J.S. Paintal in 1966. They were printed in Survey of India's 103 (P.Z.O.) Printing Group located in Dehra Dun. The tradition of bringing out reports and manuals continued. Hence, we find publications (sometimes mimeographs) on map projection, different aspects of geodesy, important meeting minutes and the like. Recently, Technical Paper No. 01/2020 on the Development of Geoid Model for Uttar Pradesh & Part of Uttarakhand under National Hydrology Project (NHP) was brought by the Director of the Geodetic & Research Branch (G&RB). However, with the introduction of computer-based techniques, digital disruption took place. The earlier manuals and handbooks became less relevant. They were taken over by software-based technologies which keeps on changing with every new version.

Writings on Survey of India was not always an in-house activity. Several authors were attracted to the then biggest scientific activities under the aegis of Survey of India. Attempts were made to document the events however the objectives and intentions were different. Some of the authors had close connections with the Survey either due to army or colonial background. This is apparent from the book by Jerry Brotton (2012) on *A History of the World in 12 Maps* (p. 344).

"In contrast to the Ordinance Survey's difficulty in providing standardized maps of England's complex and entrenched system of land ownership and management, the English East India Company assumed it would be much easier to survey the overseas possessions like India by using new scientific techniques and simply ignoring the local methods of mapping and owning land, notwithstanding the country's size. In the 1760s the company began providing financial support to individuals like Rennel for surveys that cumulated in the Great Trigonometrical Survey of India. The survey was judged complete by 1843, but work carried on decades, and, like Cassini surveys, it has no decisive terminal date. In the words of Matthew Edney, the survey's most distinguished historians, the surveyors 'did not map the "real" India. They mapped the India they perceived and that they governed', and as a consequence created 'a British India.'"

Reference has been made to the book entitled *Mapping an Empire: The Geographical Construction of British India, 1765-1843* by Matthew Edney (1997). The colonial inclination of both the books under reference has been quite apparent. The interest of documenting the activities of Survey of India continued after independence as well. The R.H. Phillimore's five volumes on *Historical Records of Survey of India* are classic example. However, they cover a limited period. He described survey operations in different parts and stretches of undivided India. The five volumes provide details about marine surveys, revenue surveys, Pindari and Maratha wars, military routes, methods of surveys, details of great trigonometrical surveys, astronomical observations, natural history, metrological observations, instruments, projections and scales and the like. They include coloured maps, pictures, sketches and biographical notes of the surveyors. The whole operation appears to be a semi-military in nature involving army officers. Phillimore in the Preface of volume III made following observations:

> "Of the five great surveyors Mackenzie was the only Engineer, Lambton and Hodgson were Infantry officers, Blacker a Cavalry man, and Everest a Gunner. It is not generally recognized how few of the early surveyors of India came from corps of Engineers."

Mackenzie was the first Surveyor General of Madras and later of India as well; and 'pioneer of topographical surveys. Phillimore was himself a Colonel in the Royal Engineers and Survey of India. The volumes by him are considered to be an authentic account of the Indian surveys. Though documents do exist, the befitting historical account of 19th century onwards had not been attempted. However, on the occasion of bicentenary of Survey of India, under the guidance of Lt. General G.C. Agarwal, the then Surveyor General of India, *A brief History of Survey of India, 1767-1967* was brought out. This mimeographed volume contained the important features during different historical periods. It includes narration about the beginning of scientific surveys and challenges of post-independence India. Further, in 1990, a souvenir was brought out by the Survey of India on the occasion of birth bicentenary of Col. Sir George Everest. Lt. General S.M. Chadha was then the Surveyor General of India. This was also a mimeographed volume centered around the contributions of Col. Everest. Recently, K.G. Behl brought a poetic account in his book entitled, *Glimpses from Survey of India: Covering 250 Years.* Further, what Suresh Prabhu and Shobhit Mathur

(2003) wrote for educational institutions is true for scientific establishments like Survey of India as well:

"A biography of an educational institution additionally serves as a model for other institutions, highlighting best practices and successful strategies that can be applied elsewhere. It could show the value risks and trying new things, which can encourage younger institution builders to do the same."

Inspired by the authors who have contributed towards documenting, recording and publishing the narrative of the Indian survey department, I mooted the idea of writing this *Memoir of an Indian Surveyor General*. This will at least cover the period of which I was with the Surveys. I have been associated with Survey of India in different capacities. Before joining this department, I was associated with different committees as a representative of another national level mapping institution, *i.e.* the National Atlas & Thematic Mapping Organisation or NATMO. Common strategies for preparing the District Planning Map Series (DPMS) or commercialization of maps were developed. I was entrusted with the responsibility of the Surveyor General of India three times: (a) December 2001 – February 2005, (b) July 2007, and (c) July 2008 – October 2008. Later, I have been chairmen and member of different Union Department of Science & Technology (DST) committees which have directly or indirectly links with the Survey of India. Some of these committees were National Geospatial Programme (erstwhile Natural Resources Data Management System), National Spatial Data Infrastructure, Survey Board, Reorganization of Survey of India, National Geospatial Policy, selection of mapping software for Survey of India, Central Survey Programming and Planning Board, Empowered Committee for Survey of India and the like. Hence, this memoir will have the reflections of my above-mentioned associations with the Survey for over 25 years. As a result, I may be considered to be insider or even outsider of the system. In both capacities, there are certain advantageous and limitations as well.

It is not the first time a treatise has been attempted related to the work of Survey of India; or a book on scientific or technical matters written by a Surveyor General. Charles E.D. Black in 1891 brought out *Memoirs on the Indian Surveyors 1875-1890* covering a period of 15 years only. It was published by the order of Her Majesty's Secretary of India in Council,

London. Further, Sir Clements Robert Markhan brought out two books entitled (a) *A Memoir on the Indian Surveyors*, published by W.H. Allen & Company in 1871 (2ⁿᵈ edition in 1878); and (b) *Major James Rennel and the Rise of Modern English Geography*, published by Cassell & Company in 1895. Even Sir George Everest brought out following books describing measurement of the arc:

1. *An account of the measurement of an arc of the meridian between the parallels of 18⁰ 3' and 24⁰ 7' : being a continuation of the grand meridional arc of India as detailed by the late Lieut.-Col. Lambton in 1830.*

2. *An account of the measurement of two sections of the meridional arc of India: bounded by the parallels of 18° 3' 15"; 24° 7' 11"; & 29° 30'48" in 1847.*

On the occasion of the bicentenary functions, several publications were brought out. One was *The Great Arc 200 Years: Celebrating the Quest – A Saga of the Agony, Passion and Triumph of Precision Inch by Inch along the 78ᵗʰ Meridian* (Nag 2002). It was published by the Union Department of Science & Technology. A coffee-table book on the *Great Arc* was also produced to mark the occasion. In addition, several pamphlets, booklets, souvenirs and other publications were brought out to observe this event. All these activities motivated me to write the memoirs in order to records the major events which took place during my involvement with Survey of India. Nevertheless, I am still associated with this great organization, both at individual and institutional levels.

It is about two decades when I took up the responsibility of a Surveyor General of India. Over the years I have collected materials to write this memoir. They are in the form of copy of notes, meeting minutes, lectures, published research papers, interviews to the media, power point presentations, tender documents, entries in dairies, printed and published materials, photographs, and correspondences. Some are in the form of hard and soft copies. As a result, I have to go through thousands of pages, compact disks, pen drives, e-mails and sometimes old floppies as well. Nevertheless, all the desired materials were not always with me. Hence, I have to take help of the offices of the Survey of India, particularly Survey General's Office (SGO), Geodetic and Research Branch – both located in Dehra Dun, and its office in Kolkata, mainly the libraries of these offices. I had additional advantage of being in Kolkata because the earlier headquarters of the Survey was located in this city. It was a pleasure to get cooperation the officers of this office. Furthermore, due to corona pandemic, I got the advantage of getting ample time to concentrate in writing this Memoir.

I am particularly indebted to Dr S. Chandrasekhar, Secretary to the Government of India, Department of Science & Technology for encouraging me in this pursuit. He kindly agreed to write the Foreword of this volume. I take this opportunity to thank Prof K.M. Pandey, Head of Department of English, Banaras Hindu University for going through the manuscript and giving valuable suggestions.

The whole objective of this memoir is to record the scientific and technical initiatives taken during my tenure so that they are not overshadowed by other activities taken during the same period. Some of such initiatives lead to the establishing of National Spatial Data Infrastructure (NSDI) and National Geo-technical Facilities (NGF). Both were initially established as a part of Survey of India. NSDI, based in New Delhi, was taken over by the Department of Science & Technology having close cooperation with its National Geospatial Programme. NGF, the latter one, came under the control of other institutions and finally it has been reverted to the G&RB Directorate of the Survey of India. The other initiatives were conversion of topographical details on colour separated on thick glass plate to digital data; transformation of field operations, introduction of remote sensing in a limited way; geodesy, large scale mapping, implementation of dual map series; reorganization of the whole setup spread all over the country; establishment of modern printing facilities and special cell for the Ministry of External Affairs (MEA) in Palam, New Delhi; creation of survey office (later GDC) in Chennai for Tamil Nadu; construction of buildings in Jaipur and Gandhinagar; development of SGO GIS Cell and the like. Implementation of all such enterprises was not easy in an organization which is known for its heritage and traditions.

This volume is likely to demonstrate how a 250-year-old setup has been transformed to a modern mapping, geospatial and survey organisation. The whole process may become a role-model for other countries undergoing similar processes of change with continuity. Each major initiative has been dealt as a separate chapter in this book.

Dr Prithvish Nag, May 2023

Chapter 1

Introduction

"The old order changeth yielding place to new".

Lord Tennyson in The Passing of King Arthur

Cartography is about maps. This includes the art, science and technology of map making. Cartography has always been closely associated with geography and surveying. The early mapping may be associated with the development of symbolization (or presentation) method that was used by early mapmakers. The definitions tend to emphasize map-making while recent descriptions also include map use within the scope of cartography. The history of Survey of India (SOI) has been a part of the saga of nation building. It was never realized that the foundation of modern India will become synonymous with the activities of this department, but its contribution has never been emphasized – not even by the SOI itself. How scientific foundation or development initiatives could have been possible without the anticipatory actions taken by the department? The process of understanding of the priorities in development and defence has been the forerunner in all respect.

All such developments were of course not without price and efforts. The scientific measurement of the country had several ramifications. Surveyors traversed from regions to regions, waiting for an opportune time free from natural, man-made and logistic problems in order to pursue their task. Hazards, diseases, snakebites, dacoity, hostile local people, battles and the like came in the way of their mission. Surveyors penetrated the jungles, climbed mountains, crossed rivers and fixed poles, stations and control points throughout the length and width of the country. There was no respite whether it may be on the slopes of Western Ghats, or swampy areas of Sundarbans, or ponds and tanks, ox-bow lakes and meandering rivers of Bengal, Madurai or Ganga basin. The deserts, mighty Himalayas, Rann, River Chambal or Gandak, *terai* or *dooars* were also not spared. The survey has been associated with the waves of Arabian Sea or Bay of Bengal, dust storms of Rajasthan, cyclones of Eastern Coast, cold waves of the north and the widespread monsoons and heat all over the country.

It was against this price, determination and missionary zeal that the foundation of the country was laid. The knowledge assets built over years and generated with whatever technology then available proved to be valuable. The best part of the whole process is that a new information set which may be based on latest technology like aerial photography or global positioning system takes benefits of the earlier generated information. No piece of information is wasted and has relevance even after decades. How such an empire of knowledge has been built up? It is not only a technological feat that is most important but it is an emblem of pursuance of one of the longest scientific experiments carried out in the world *i.e.* mapping the nation against all odds. The legacy and traditions continue. It is a matter of great interest to know how such a super structure of information has been built. How this great scientific experiment has been carried out for so long. What were the compulsions and apparent benefits? Why the colonial rulers and later the independent government with limited resources continued their interest in this expensive exercise? These are some the questions that will keep on bothering the people interested in the history of science of this country.

The science of map-making could have not been possible without the knowledge of shape and size of the earth. This was required for determining the coordinate of points on the earth's surface, which have been used for map-making. Efforts were made from the very beginning by various philosophers and scientists for determining shape and size of the earth. Man's first conception of the earth's shape was a plane. Greek philosopher thought it a sphere. Sir Issac Newton pointed out that it must be spheroid. However, development of science of surveying and mapping of earth's surface in an orderly manner has been ascribed to Eratosthenes (276-198 BC) who based the compilation of his comprehensive map of the world upon his famous determination of the circumference of the earth. Claudius Ptolemy, a mathematician, astronomer and geographer produced a map of India in the 2nd century AD. However, the golden age of Indian renaissance started when in 5th century, genius Aryabhat calculated the earth circumference to be 25,080 miles and wrote *Surya Sidhanta*. The next contributor to Indian geography were Chinese and Arab travellers who left the interesting and detailed accounts of their journey. It is not wonder that this mystic land attracted so many adventurers.

The basic concepts of map-making *i.e.* scale, generalization of features etc. were known in India as evidenced in *Purana*. Various references in the *Vedas* testify to lead to this conclusion. It has been established that art of surveying technique of mensuration of areas were well developed in ancient India; the manual known as "*Sulva Sutra*" (science of mensuration). In *Vedic* literature *sutras* (formulae) were provided for the measurement of angles, distances and areas. Our ancient knowledge of astronomy, astrology, rocketry and geography is well known and has been apparent to Indian scholars that the *Vedas* and later Sanskrit works contain these truths. In *Vedas, Upanishads,* the great epics *Ramayana* and *Mahabharata, Puranas, Manusmriti,* works of Panini, Patanjali and Kautilya and poem of Kalidasa, there has been reflection of geographical accounts of India. The Vedic concept of motherland, mentioned in *Prithvi Sukta* in the *Atharvaveda,* has no parallel in any other literature written outside India. The Indian astronomers and mathematicians like Aryabhat, Baraha-Mihir and Bhaskaracharya discovered several truths like shape of the earth, its rotation around the sun, and even the force of gravitation.

Although, geographical knowledge traces its historical background from very beginning of civilization, but that has been the domain of history. Knowledge of maps as a science grew with time and religious thinkers also started delving into its mysteries. The age of great exploration started with the quest for uncorrupted truth. The major exploration got underway at the end of fifteenth century with Columbus crossing the Atlantic in 1492, Vasco da Gama circumnavigated Africa in 1497 and the Magellan's expeditions circumnavigated the world between 1519-1522. The Dutchman Snell (1591-1626) carried out the first measurement of angles and distances and made the rigorous study of refraction. The French Clergyman Picard, in 1670 made the measurement of size of earth. His results of 6,275 km for the radius of earth were the first improvement on Eratosthenes determination of the size of the earth. Newton's theory, that the earth should be oblate because of the centrifugal force, caused by the spin was validated by the measurements of two meridian arcs, one at the equator and the other closer to pole measured by two survey expeditions organized by French Academy of Sciences.

The dissemination of geographical knowledge was limited till 15ᵗʰ century before the invent of printing technique due to the labour and skill involved in duplicating graphic data. Once, however, lithographic techniques were discovered map-making got a stimulus, particularly in Europe where adventurous men were out to conquer new worlds for religion and commerce. Father Monserrate, a Jesuit Missionary, brought out a map after a visit to Akbar's court. This was a first of those maps which are based on measured routes and astronomical observations. Sher Shah Suri and Todar Mal's revenue maps, based on regular land survey systems, were well known in the medieval period and continued to be in practice during the mid-eighteenth century. These are available information of the surveys instituted by Akbar during 16ᵗʰ century; measurements being made by a hempen rope which was replaced by a *"Jarib"* of bamboos joined by iron rings. The other noted cartographers of the time also published their vision of maps of India, but in 1752 the French geographer, Jean-Baptiste Bougigon d'Anville published map of India and put Indian geographical knowledge on a scientific footing. Further, the pioneering work done by French expeditions showed that terrestrial measurements (angles & distances) were the main tools for the task of relative positioning. The technique of triangulation, astronomical determination of positions and azimuths as well as leveling were started mid seventeenth century.

In the early days of the imperial powers in India, the geographical knowledge of southern parts of the country was greatly increased during the period of wars of political supremacy but northern parts of the country (Bengal, Uttar Pradesh etc.) was sketchy. Nevertheless, attempts were made from very early times to establish the geographical locations of important places. Raja Jai Singh Sawai of Jaipur of pre-British period established astronomical observatories at Jaipur, Delhi, Mathura, Ujjain and Varanasi and determined geographical positions of various locations.

During the middle of 18ᵗʰ century, latitude could be easily determined by observing the meridian altitude of the sun or stars. For longitude determination, one had to wait for favourable phenomenon such as an eclipse of sun, moon or satellites in clear sky with similar observations to be made at some known places. In 1787, astronomical observations for latitudes and longitudes at various places were made covering India. The proper astronomical centres were established in due course of time using sextants, chronometers and telescopes. The results of these instruments were imperfect because of their accuracy level and faulty mathematical tables. It is interesting to note that because of difficulty in determining longitude, early maps of India did not indicate longitude with reference to Greenwich meridian but are referred to arbitrary meridian passing through Madras or Calcutta.

Most of the early maps were based on local surveys carried out by cursory methods. In 1776, comprehensive instructions for preparing the maps on the scale two mile to an inch formulated: in which distances were measured by perambulator rather with chains and bearing to the conspicuous hills. Short base lines were laid and measured and distant points were fixed by triangulation. According to Murli Manohar Joshi (Appendix VII):

"In hindsight, the Great Arc odyssey is capable of being viewed in many different ways and at many levels. Conventionally, it has been seen as the mapping of an empire - a typically grandiose, colonial enterprise to consolidate the military, economic and territorial gains of an imperial state. Scratching beneath the surface, we begin to see the magnificent obsession of a few individuals, using the cloak of imperialistic ends to justify their fascination with a particular branch of science and an almost fetishistic passion for measurement and empirical observation. If we go a little deeper we can see the unfolding of a still larger, grander, story - that of a unique partnership of minds. We need to see through the colonial context and the asymmetry of political power at that time, to appreciate the collaborative nature of the project and the strength of the team effort which went into it. The Survey was not a military campaign with star British generals leading an army of passive Indian subalterns and foot soldiers. Nor was it an industrial enterprise which could employ armies of cheap Indian wage labourers regimented into an assembly-line order. The utilitarian ends of the exercise were not apparent enough. There was no pot of gold or buried treasure at the end of it. Yet, thousands and thousands of men offered their skills, their brainpower, their brawn, their patience and indeed their lives, for a scientific pursuit."

Historical Background

The Survey of India traces its birth starting from Major James Rennell appointed as Surveyor General of Bengal on the 1st of January 1767. In those days, there was an urgent need for having a picture of the country showing general course of main rivers and the location of principal towns. This task was taken up with speed, and in the process, produced some serviceable maps of areas of Bengal and Bihar in less than twelve years. These maps, however, could lay no claim to accuracy of details but were sufficient to meet the needs of the time. Rennell also produced *"Map of Hindoostan"* in 1783 after relinquishing the post of Surveyor General.

Till the beginning of the 19th century, the progress of topographical surveys in Madras and Bombay presidencies was more or less independent and uncoordinated with the work of Bengal presidency. This was not a satisfactory situation and retarded the progress considerably. It was only in 1787 that an accurate survey of the seacoast from Madras to southernmost extremity of the peninsula was taken up by running a 300-mile line of triangles along the coast with a view not only to ascertain the actual line of the seacoast but a complete survey of peninsular of India. This survey was the first Indian survey based on triangulation.

Towards the close of 18th century, theodolites now considered primitive had been brought in use. The angles and bearings were measured with theodolite and pocket compass for direction of the road etc. The technique of plane tabling was first used in 1792. Plane tabling survey was subsequently developed into an art and has been extensively used down to present time for topographical surveys in all types of terrain. Even now, though the modern techniques of surveying are coming up, yet this simple technique has been widely used in various surveys such as large-scale mapping, engineering survey, cadastral survey and the like.

The period of piecemeal surveys was over by end of 18th century. A new era of William Lambton and George Everest began that was a coordinated consolidation. The foundation of a truly scientific Survey of India was laid which was the beginning of a stupendous work and occupied the lifetime score of noble and devoted surveyors. A network of primary triangles was established by the trigonometrical surveys. It was a magnificent scheme,

timely conceived and brilliantly executed. Although, technique of triangulation, astronomical determination of positioning and azimuths as well
as levelling were initiated in mid seventeenth century, but the scientific
procedures started only by the end of 18th century when a project for the
measurement of an arc of the meridian through a network of trigonometrical surveys covering the Indian peninsula was mooted.

The actual work of the great trigonometrical survey was commenced on
10th April 1802 by the measurement of a baseline near Madras. This baseline was established using steel chain which consists of 40 links of 2 1/2
feet each, measuring in the whole 100 feet. This baseline was measured
with the aid of coffers (long boxes) as it was required for the triangulation of the *Great Arc* where utmost possible accuracy was aimed at. From
this baseline, measurement of a series of triangles was carried up to the
Mysore and the second base was measured near Bangalore in 1804. To
start with primary reference station of origin was the Astronomical Observatory at Madras (now Chennai). Having connected the two sides of the
peninsula, Lambton devoted much of his labour to the measurement of
an *arc of meridian*. The series measured for the purpose is known as the
Great Arc Series. In addition to the measurements of these series, webs of
triangle extended for establishing the positions of main cities. This idea
of web of triangles was replaced due to cost effectiveness by an all-India
grid composed of criss-crossing 'chains' or 'bars' of triangles centred on
the *Great Arc*. The holes on the grid could be filled later by cheaper and
less rigorous topographical surveys. This idea gave birth of the name
"grid-iron" (Appendix II).

The *grid-iron* layout consisted generally of an outer frame of two extreme
meridional and two longitudinal series closing at each junction on
a measured baseline. In order to bring the *Great Arc* across the plains,
masonry tower stations were built which were about 50 feet high. The
first essentials of every observation station, whether on hilltop or tower, or
otherwise was the stability of the instrument and immovability of the mark
over which instrument and signals were centred. Since these marks were
established after putting hard work, thereafter, they were handed over to
the civil authorities when all connections were completed. In 1866, it was
ordered that all stations of *Great Trigonometrical Survey* should be placed
under the official protection of district magistrates and were visited peri-

odically. This practice is still in vogue till the present time for all primary *Great Theodolite* (GT) stations and benchmarks.

Lambton main instrument referred as the GT which was a marvel workmanship of those days. The horizontal circle was 36 inches in diameter and the vertical circle 18 inches; each was read by two microscopes. This theodolite was used by him and his assistants and then by Everest and others till 1866. Various other theodolites were used for observations for meridional series *viz.* 36-inch theodolite, which was built up from Lambton great theodolite, and 34-inch, 24-inch, 18 inch and 15-inch theodolites. For laying out the series and running secondary and minor triangulations, small theodolites were used *i.e.* 14″, 12″ and 7″. For baseline measurements, compensation bars and other baseline apparatus were used. The compensation bars remained, however, the only means available for measuring the baseline of the main triangulation framework - Vizagapatnam and Cape Comorin base were measured during 1862 to 1869. In 1856 standard yard arrived from England and the following year a special room at Dehra Dun was set aside where subsidiary standard could be laid off or compared by microscope as and when marking of staves for levelling operations required. Standard spirit levels were used during those days.

Triangulation or levelling computation were made to a routinely, adapting rules and formulae of the department. Computation forms were lithographed at Calcutta under the direction of the chief computer. One of the greatest contributions which Radhanath Sickdhar, chief computer, made to the *Great Trigonometrical Survey* was the preparation and publication of a set of tables to be used with departmental formulae, and the computation forms. The first official lists of geographical coordinates were published in 1842; and the first edition, entitled *Tables to facilitate the computation of a Trigonometrical Survey and the Projection of Maps,* was published in 1851. Further, auxiliary tables to facilitate the calculations were published in 1868.

For the dispersal of triangular error, method followed by Everest was tested against new method devised by Gauss. Radhanath Sickdhar first tried two simple figures and obtained results closely agreeing with old method. He tried this on complicated figures that occur in the trigonometrical survey. The results were highly satisfactory, showing that the greatest discrepancy between Gauss's and Everest's methods would not exceed 0.14 arc seconds.

The final distribution of errors and reduction of results of the triangulation of the *Great Trigonometrical Survey* was carried out under the direction of General James Walker whilst he was superintendent of the trigonometrical survey. For this purpose the whole triangulation was divided into five zones – northwest, northeast, southwest, southeast quadrilaterals and the southern trigon.

Lambton and Everest did not go deeply into the subject of heights above sea level. Lambton first connected to the sea at Madras in 1802, but for his great Central Arc he preferred the connection made at Cape Comorin in 1809. From this he brought up his height by vertical angles from station to station. Surveyors in India had no professional interest in the measurement of the vertical rise and fall of tides along the coast except to find the level of the sea from which to calculate their land heights. Lambton followed the practice of his time in calculating his heights from low water. It was only in 1837, when the mean between high and low tide observations for at least for half a month were proved to agree closely from one place to another. Self-registering tide gauges were invented during 1830-33 and was established at Colaba observatory, Bombay (now Mumbai) in 1842. Tide gauges at various other places were also connected through spirit level during 1851 to 1860. Leveling was initiated on scientific lines in 1858. It was General Walker who was the founder of this activity and bestowed upon the subject before he initiated the field work.

Waugh took over in 1843 from Everest and during seventeen years of his administration, triangulation series was advanced eastward to over Ganga valley to Calcutta and extended up to Assam. Regular observations to the Himalayan peaks were taken. He was responsible for the discovery of the highest mountain in the world, 29002 feet above the sea, which he recommended should be named after George Everest who had built up the triangulation system by which the discovery was made possible (C&Q 16).

Regular astronomical observations for azimuth and meridian were continued along both meridional and longitudinal chains of triangles as check against accumulation of errors in different directions. Both, Lambton and Everest, had been well aware that their observations were influenced by visible mountain masses and variation of density. Various other mathematician and geodesists worked on this subject and attracted wide attention suggesting the value of pendulum observations for the determination of variations of gravity and introduced the *theory of compensation or isostacy*.

In the recorded history it has been mentioned that the Indians were being employed for menial jobs, carrying instruments and equipment, clearing obstacles in the line of sight, jungle clearing, erect hill stations and take Instruments and equipment to the tops, sending signals with flags, making measurements with chains and help in spreading baseline. Those who learnt English were tried to act as *munshi* or writer to record details of tours. Precision work of observations was being done by the British officers and their European retinue till 1830. Major George Everest, who later succeeded Colonel William Lambton, introduced a lot of changes in field operations and was a creative and innovative genius and his introduction of compensation bars for measuring bases, heliotrope flashes and lamps for signals for observing stations, method of ray tracing, erecting flag staffs, designing of survey towers and the like were unique but needed more trained Indians to manoeuvre those. Though initially reluctant but started inducting more and more Indians as the work expanded and his faith in them grew. Lord William Bentinck, the Governor General, in 1833 decided to encourage the employment of educated Indians in their service. About this whole operation, Murli Manohar Joshi mentions on the occasion of India Day in London in 2004 (Appendix VIII):

> "One, that the spirit of conscious scientific enquiry is a part of every segment of Indian society and is not limited to those who are fortunate enough to attend a University course in it. That is why we could have a lowly clerk like Ramanujan dedicate his total life to the pursuit of mathematical abstractions. A Radhanath Sikdar who became George Everest's, Chief Computer and without whom the Great Arc could literally not have achieved the heights of Mount Everest. A Syed Mir Mohsin Hussain - a watchmaker who went on to engineer, on his own, the most complex mathematical instruments, some of which are on display here. The *pundits* - Nain Singh and Kishen Singh - ordinary school teachers in a vernacular tradition who became pioneering surveyors and geographers."

Meridional Chains & Arc

The measurement of *Great Arc* from Cape Comorin to Himalayas was completed by 1843. The *'grid iron'* system consists of meridional chains of triangle tied together at upper and lower ends by longitudinal chains. Ambitious scheme of triangulation commenced with the Great Arc Series, having Dehra Dun base at the north and Sironj base in central India as the southern end. From Dehra Dun base northwestern Himalayan series was extended while from Sironj base the Calcutta longitudinal series was extend up to Karachi base. In 1887, observation of a series from north-western Himalayan known as Kashmir series was started. The height of stations averaged 17,000 ft in this series. The second highest peak, next to Mount Everest, was found during this triangulation which was 28,290 ft high and was named K2.

The *Great Trigonometrical Survey Triangulation Network* of India and adjacent countries was started in the year 1802 and by about 1880, several triangulation series had been observed to warrant their simultaneous adjustments. This triangulation network was first adjusted to form a self-consistent whole in 1880. Adjustment of this horizontal network was based on Indian Geodetic Datum. Everest ellipsoid was used as reference surface in India. This surface has been named after Sir George Everest. A reference ellipsoid was defined by various components *viz.* semi-major axis (a), flattening (f) and coordinates *viz* latitude, longitude and deflections of the vertical: meridional and prime vertical, and geoidal undulation at the origin. Everest adopted Kalianpur in the central India as origin. Various components of Everest ellipsoid or spheroid and its orientation at origin worked out in piecemeal manner in various campaigns.

In the year 1937, another adjustment was attempted, incorporating the new triangulation series observed after 1880, the LaPlace stations to control directions and new baselines measured with wires between the year 1930 to 1934 to control the scale of the triangulation series. In this adjustment, instead of resolving the simultaneous normal equations formed after incorporating new data, graphical technique of adjustment was employed. This technique was not considered appropriate, and the adjustments found no practical use. Nevertheless, the adjustment of 1880 has remained the basis of all India triangulation and mapping except for a constant change of –

(2'27".18) in longitude. The mapping activities were scientifically executed and based on triangulation series adjusted in 1880. Most of the maps were on the scale half inch or larger. This change was essential as longitude of Madras observatory was revised in 1905. One of the objectives of the above enterprise was to prepare an *Atlas of India* to be published by the East India Company. According to the *Scientific Intelligence* journal of 1829 (p. 347) (Q&C 16):

> "This noble work, of itself a splendid monument of the munifi-cence of the East India Company, is upon a scale of four miles to an inch, and takes from actual surveys, which when completed, will for a map of India on one uniform plan. The project was first conceived by Colonel McKenzie, and a large portion of those parts already published were surveyed under his superintendence. The surveys on the northern parts of the peninsula have for their basis the triangulation of Colonel Lambton, who extended a set of prin-cipal and secondary triangles over the whole country.
>
> The sheets are published as they are completed; some of them have blank spaces, to be filled up as the surveys proceed; nothing allowed to go forth to the world which is not founded upon actual survey."

Mapping activities opened new possibilities with the introduction of aerial photographs during early years of war. The science of photogrammetry had not developed but their immense value impressed the surveyors. The aerial survey techniques were however not much used in the department up to 1939 except for training for military survey purpose and for survey of inaccessible areas. The primary map scale 1 inch to a mile was adopted in Survey of India in 1905. It was also decided that these maps may be produced in colours and show the relief by rigorous contouring. The requirement of printed maps increased during 1939. The main printing organization at Calcutta (now Kolkata) could not meet the mapping requirement. Therefore, the present site at Dehra Dun was selected and the printing machines were installed and by middle of 1943, the printing group was in operation with three large high-speed machines. At the time of independence (1947) about 60 per cent of the country was covered by primary scale of mapping on 1 inch to 1 mile.

Post-independence Developments

During 1947 onwards, there was an immense increase in the developmental activities which continue in one form or other. SOI was more entrusted with the development schemes considering the post-independence reconstructions priorities. The normal topographic survey became secondary.

During the first and second five-year plan, the department faced with demands for survey which required for more survey potential than it had. During the first five-year plan (1951-56), nearly 70 per cent of the department's potential was employed in developmental surveys and in second five-year plan about 60 per cent potential could be directed for this purpose. The department had diverted workload with the switching over to metric system in 1956, the basic map scale having been changed to 1:50,000 from 1 inch to 1 mile. It was during second five-year plan, when advancement in various techniques of mapping were available and there was pressing demands for maps for the development of the country. Survey of India, keeping pace with modern technologies of surveying and mapping continue to adopt various activities in the field of geodesy and geophysics, topography, photogrammetry, cartography, printing and manpower development.

Map-making has evolved in response to theoretical developments, technological advancements and changes in society's information needs. With the advancement of technology, mapping technology has undergone major changes. Experiments with photography, aerial photography and remote sensing then added new dimensions to the methods of map making and reduce the time involved in the production of a map. Topographic maps were produced using these technologies with ground control points to ensure positional accuracy. However, not all maps were concerned with location of physical features. Thematic mapping expanded rapidly following the introduction of censuses and other surveys which were taking into consideration of extensive demographic and socio-economic data.

The past two decades have seen dramatic changes in cartography due to the development in digital and communication technologies. Earth observation satellites now provide coverage of earth surface at a variety of spatial resolution ranges from few meters to several kilometres. Global

Positioning System (GPS) allow precise determination of horizontal and vertical positions. Computer mapping systems have advanced in the form of Geographical Information System (GIS) which are now widely used in planning, resource management and facilities management applications. Development of desktop mapping technology and internet access to electronic data sets had made powerful map-making technology available to anyone having a personnel computer.

Mapping embraces production of various topographical, engineering, geographical, thematic and earth science and research maps, which are essential tools for planning and developmental activities. Cartographic products have been the base material for planned development, sound administration, national security and military operation, management of natural resources, peaceful coexistence and every facet of nationwide activities. The defence forces, planners and other law and order enforcement agencies, economists, geologist and all other working in creative sciences find a map an indispensable tool.

Large scale maps are being prepared in different parts of the country for developmental projects *viz.* hydroelectric projects, irrigation projects, industries establishments, canal area development, flood plain zoning and other various developmental activities. Tunnel alignments, underground power houses, monitoring geometrical shapes of various structures require accurate determination of coordinates in plan and height which are prerequisite for any such activity. It is imperative to say that cartography and planning must be properly interlinked and have to be dovetailed for scientific developments of a region.

All the developmental activities require prior mapping. The ecological complexities of the environment can be assessed only with periodic preparation of maps of the concerned areas. A topographical map of region depicting its landforms, drainage, coastal features, vegetation, communication and other detailed distributions provides the knowledge of the relationships of various factors necessary to plan and carryout intensive development activities effectively. With sophistication in the field of aviation, communication, meteorology, hydrology, forestry, tourism, urban and rural development, environmental planning and education, the demand for cartographic products have multiplies over the year.

Existing Setup

Survey of India in its assigned role as the nation's principal mapping agency continues to bear a special responsibility to ensure that the country's domain is explored and mapped suitably to provide base maps for expeditious and integrated development; and ensure that all resources contribute their full measure to the progress, prosperity and security of the country for now and for generations to come as well. Further, it continues to function as adviser to the government on all survey matters, *viz* geodesy, photogrammetry, mapping and map reproduction. However, the main duties and responsibilities of SOI have been enumerated below:

- All geodetic control (horizontal and vertical) and geodetic surveys

- All topographical control, surveys and mapping within India.

- Mapping and production of geographical maps and aeronautical charts.

- Surveys for developmental projects.

- Survey of forests, cantonments, large sale city surveys, guide maps and the like.

- Survey and mapping of special maps.

- Spellings of geographical names.

- Demarcation of the external boundaries of the country and advice on the demarcation of inter-state boundaries.

- Training of officers and staff for human resource development within the department, trainees from central or state governments and from other countries.

- Research and development in cartography, printing, geodesy, photogrammetry, topographical surveys and indigenisation.

- Gravity surveys.

Apart from the above responsibilities, it has been entrusted to make predictions of tides at 44 ports including 14 foreign ports; and publication of *Tide Tables* one year in advance in order to support navigational activities.

Organisation Setup

At the time of the beginning of the millennium (2001 AD), SOI had a traditional setup keeping in view of the administrative, technical, and field requirements. There were four zones and 18 directorates out of which seven were specialized directorates. In order to carry out field surveys, there were 58 fields units or *'topo parties'* spread all over the countries and attached to different directorates. Apart from the field data, another source of ground information was aerial photography. Large dependence on this source could be understood by the fact that there were 17 'photo units'. In addition, there have been specialized units in different directorates such as (a) Geodetic & Research Branch, (b) Directorate, Survey Air, and (c) Survey Training Institute. Later, other directorates were added for meeting the requirements of the modern technology. Map printing was a major activity then in the SOI. Such maps were required by the security forces. Hence, it had five printing groups located in Dehra Dun (2), Kolkata, Hyderabad and New Delhi (vide Chapter 6).

Printed Products

As mentioned earlier, the primary role of SOI had been the production of topographical maps (Fig 1.1) at standard scales (Table 1.1):

Scale	Contour interval	No. of sheets
1:250,000	100m/200m in hills	394
1:50,000	10m/20m in hills	5,104
1:25,000	5m/10m in hills	19,540

Table 1.1 *Topographical scales*

Figure 1.1 *Example of topographical map*

SOI has also been publishing other popular maps. These maps are considered most authentic and become base maps for all types of small-scale maps concerning India. External boundaries have been correctly depicted in such maps. In addition to the topographical maps mentioned above, there are following popular geographical maps:

- General Wall Maps

- India & Adjacent Countries on various scales

- Political Map of India, Scale 1:4 M

- Physical Map of India, Scale 1:4.5 M

- Railway Map of India, Scale 1:3.5 M

- Road Map of India, Scale 1:2.5 M

- The World Map, Scale 1:20 M

- Children's Map of India, Scale 1:4 M

- National Parks in India, Scale 1:5 M

- State Maps on 1:1 M Scale

- Special Maps

- Forest Maps

- Antique Maps

- Tourist Maps

- Trekking Maps

- Guide Map Series

- Discover India Series

- Plastic Relief Maps

- District Planning Maps

- Maps for Civil Aviation Organi-sation on 1:1 M scale

- World Aeronautical Charts and International Maps of the World (IMW)

Other Services

SOI has been involved in human resources development as well. Its Survey Training Institute is located at Hyderabad continues to impart training for technicians, technologists and professionals. However, its name has changed. The Institute runs basic, mid-career, graduate and advanced post-graduate level, management and users-oriented courses in all aspects of surveying and mapping. Training facilities are also been extended to neighbouring countries. Further, following geodetic and allied geophysical activities are extended by the Geodetic & Research Branch or G&RB of SOI:

- Establishment of primary planimetric and height control for national control framework and mapping.

- Geophysical surveys including gravimetric, geomagnetic, astronomi-cal and tidal surveys.

- Provision of geodetic data for precise alignment of tunnels, power-house complex, dams, barrages etc.

- Provision of geodetic and geophysical data for crustal movement stud-ies, dam deformation studies, verticality of archaeological structures etc.

The Research and Development initiatives have been very much part of its activities. Considering the changing technological scenario, there use to be a separate directorate for this purpose which looks in the following activities:

- Development of digital data capturing device PC EK22 from photo-grammetric instruments.

- Development of DIGIMAP software for conversion of analogue stereo plotters into digital workstations.

- Pilot project in Odisha (earlier Orissa) to device a suitable Land Information System (LIS) system for cadastral mapping.

On regular basis, services were provided to the Air Force for the preparation of IAF maps, landing charts and aeronautical surveys. Further, it has also been responsible for demarcation of international boundaries.

Recent Initiatives in Survey of India

Cartography and GIS have been strongly related to each other, yet they remain fairly independent in their central objectives. Cartography is a much older field than GIS and some of its areas of interest are not well suited for GIS like visual impacts of different mapping techniques, map theory etc. Nevertheless, cartography is strongly supported by GIS like data storage and manipulation, spatial inter-relational studies, experimentation with different mapping techniques. Therefore, GIS operations do not necessarily lead to the production of maps as final products. Although, it helps in producing more user input-friendly spatial products. Updating and changing of data and maps using GIS tools is easier than using manual method. GIS is strongly connected with fields like cartography, geography, computer science, data bases, computer graphic, landscape architecture, spatial statistic, management planning and the like.

Cartography with GIS could be used in various fields *viz* national topographic mapping, navigation, property and administrative boundary, geology, geography, surveying, other social and natural sciences, defense, emergency services, marketing and other commercial, population census, public administration and others.

SOI has been meeting the requirement of topographical maps of the nation. The department provides basic horizontal and vertical control for map making and for various other developmental activities like tunnel alignment for hydroelectric projects, fixing of dam axis, powerhouse and monitoring geometry of large structures. Gravimetric and geomagnetic surveys are being carried out for geodetic and geophysical research and exploration of various earth resources. Sea level variation are being monitored for providing vertical datum and for prediction of tides. All these data are being used for quality survey and mapping in the country. The department has digital database on scale 1:250,000 for entire country. It has digital data in the form of Digital Cartographic Data Bases (DCDBs) for parliamentary constituency, assembly constituency and administrative boundaries. It has vast storehouse of numerical data relating to geodetic positioning of points and precise levelled benchmark heights as well as gravity, geomagnetic and tidal data.

Responsibilities

The charter of duties and responsibilities for the vast spectrum of activities undertaken by SOI demonstrates its commitment to be a significant contributor in India's quest for sustainable development as also in providing for the meaningful utilization of its natural resources and to take measures accordingly. In order to meet these responsibilities in a more effective manner, SOI has been engaging itself in the following major areas of scientific and administrative activities (Appendices III & IV):

- Digitization of the Survey of India topographical maps up to 1:25,000 scale and creation of Digital Cartographic Data Bases (DCDBs).

- Introduction of New Series of topographical maps.

- Establishment of National Spatial Data Infrastructure (NSDI).

- Development of GIS for all metropolises, cities and towns.

- Establishment of digital mapping centres in all directorates of Survey of India.

- Augmentation of printing potential in SOI.

- Establishment of geodetic field parties in each zone of SOI.

- Establishment of Centre for Seismotectonic Studies.

- Upgradation of photogrammetric potential in SOI.

• Introduction of modern methods of surveying and mapping.

• Reorganization of Survey of Training Institute, Hyderabad.

• Commercialization of SOI products and services.

Digital Cartography

Though, early attempts were made to initiate digital cartography in the SOI, there was hardly any remarkable progress. Nevertheless, three Digital Cartographic Centres were established during eighties with a view to create *Digital Cartographic Data Base* (DCDB) on 1:250,000, 1:50,000 and 1:25,000 scales to cater the need of various GIS applications and for quick updating of topographical maps.

SOI as a part of its activities to keep abreast with latest technology in the science of map making, developed digital map-making programmes in early eighties with a view to adopt digital technology in the country. This organisation adopted Computer Asserted Cartography (CAC) system, the *AUTOMAP* in 1981. In this system the geographical and other data were stored in digital form and processed on demand to draw the required map output by using computer operated drafting table. It has taken a giant leap forward in 1986 when integrated digital map production systems were installed at 3 locations *i.e.* Modern Cartographic Centre (MCC), Dehra Dun; Digital Mapping Centre (DMC), Dehra Dun and Digital Mapping Centre (DMC), Hyderabad. Marginal work of digitization work was also under-taken in digitization cells created in various directorates. The capacity of the digitization cells at circle level have also been augmented by equipping them with all the necessary instruments, equipment and the like. Having overcome the initial difficulties of new technology induction in early nine-ties, the department aimed to achieve flexibility in satisfying users demands by establishing structured Nations Cartographic Data Base (NCDB).

Satellite Geodesy

With the invent of GPS, it took certain steps in this regard. GPS technology was introduced in early nineties. Its receivers were being used to provide control in high hills, snow covered area and to connect outlying islands with mainland. In several training courses were conducted on this subject by the Survey Training Institute. GPS was also being used to monitor seismotectonic movements.

Development of GIS for Towns and Cities

GIS has been introduced late in the Survey. It was adopted through the digital mapping route. Nevertheless, the major unmatched strength of the Survey lies in the ability of its field personnel being able to collect data related to ground conditions like administrative boundaries, population, and water tables in a scientific and reliable manner. Many towns and metropolises of the country were considered to be beneficial if the GIS were put in place. SOI planned to undertake fieldwork and mapping of all towns having population more than 1,00,000. Then the plan was to develop GIS for individual towns on scales ranging from 1:5,000 and 1:10,000. The attributes proposed to be covered were utilities and service, information regarding educational and health services infrastructure, commercial and business activities, and few other activities.

Products & Applications

The products of SOI were in great demand despite some times being outdated. It was expected that it should come out with a programme to assure user community to revise and update its products as per the utility of the product. Nevertheless, it became possible to generate following products on various scales from the digital topographic database as by-products:

i)   Administrative boundary database up to village level.

ii) DEM of required area as per user demand.

iii) Creation of slope and aspect map.

iv) Data required for CAD.

v)   Road and rail alignment data.

vi) Land use and land planning maps.

vii) Necessary data for rural and urban development.

viii) Image maps

ix) Utility mapping.

x) Data for disaster management.

xi) Data for network communication.

xii) Updated topographical map after capturing the information from 1-metre resolution satellite imagery with or without quick verification on ground should be made available to users. A suitable note regarding non-verification of information on ground must be entered in the footnote of the sheet.

xiii) Undertaking of supply of orthophotos on user demand.

xiv) Creation of GIS as per user requirement.

xv) Updation of existing geographic maps can be easily done by extracting data from topographic database.

xvi) Preparation of trekking and other maps of visibility series of SOI by using digital database.

xvii) Preparation of parliamentary and assembly constituency maps of India.

With the approval of the *New Map Policy in 2005*, SOI planned to provide digital topographical databases for entire country on 1:50,000 scales initially in a year's time (vide Chapter 5). Topographical data or maps in digital and analogue forms on WGS 84 (World Geodetic System 84) was targeted to make them available for general public without any restriction (existing topographical maps of SOI are on Everest datum). Generation of WGS 84 map series was then initiated in the department and the first map in digital and analogue form for the public use was released in 2002. It used the state-of-the-art technology in geospatial data acquisition, management and dissemination. Airborne Laser Terrain Mapping (ALTM), GPS, Electronic Distance Measuring (EDM) instruments, digital photogrammetry, modern printing technology and computer hardware and software were being applied in map-making process. High-resolution satellite imagery *viz* SPOT, IKONOS data etc. was being used for updating of topographical maps.

SOI was poised to take quantum leap in the form of National Spatial Data Infrastructure (NSDI). The department in collaboration with sister data producing agencies planned to assume the leadership role in meeting societal need through NSDI by developing the standards for data communication, collection and exchange of data so that it is freely available for sharing at national, regional and local levels on many different platforms or hardware and software. The NSDI framework was to provide the necessary facilities and protocols to develop cooperation among the data producers and users (vide Chapter 7).

SWOT Analysis (2002)

SOI has a very long history and tradition in its work-culture and attitude towards standards and ethics. As the topographical mapping agency, which so far, was being used mainly by defence, It had very little interaction with the public. Because of this reason, it has not developed any mechanism to ascertain what was required by other stake holders: academia, industry, NGOs and private sector. While analyzing these issues, it became necessary to understand the potentials and weakness of the organization.

Strengths

(a) Network of establishment all over the country: Survey of India being the national mapping agency has a network of establishments all over the country. This enables the department to carry out the tasks in a uniform manner and provides consistent standards for surveying and mapping

(b) Data produced by the SOI is sensitively connected with national security and territorial integrity.

(c) Data produced by SOI is very important for most of the scientific activities and national development.

(d) Trained manpower: The organization has its own training establishment for imparting training to officers and staff for adopting uniform techniques in surveying and mapping.

(e) Unique spatial data provider: The map data representing the spatial data is unique and is a primary requirement for spatial data referencing known for its standards and accuracy. The geodetic, topographic and levelling data produced by the department is known for its accuracy, though lack in currency of data.

(f) Availability of strong infrastructure for training, surveying and mapping.

(g) Availability of reliable spatial data

Weakness

(a) Over dependent on government budgetary support.

(b) Less emphasis on revenue generation

(c) Revenue generated by the department cannot be used by the department for productive work.

(d) Too rigid standards, causing delay in production of data.

(e) Not possible to enforce production of specific tailored products to meet specific user requirements.

(f) Data produced is not sufficient to meet requirement of planners and administrators for micro level planning at urban rural level.

(g) Data produced is not temporally updated and hence becomes obsolete by the time it goes to the users.

(h) Too many restrictions for free access to the data

(i) No effective dovetailing of manpower criteria and product requirement criteria.

(j) Too many technical trades persons, often working against interest of each other.

(k) Depleting manpower at all levels.

(l) Too aloof from outside world, in both technological advancement and understanding user requirements. No effective mechanism for market interaction.

Opportunities

(a) Need of data felt more and more by users.

(b) The acceptability of the data because of the standards and accuracy.

(c) Availability of techniques for online updation of data using remote sensed data.

(d) Staff strength in all levels is 40 per cent of the sanctioned posts. Hence it is the right time to analyze and infuse new blood of various disciplines to ensure changed managements at all levels.

(e) Increased awareness of data users due to the applications in GIS environment, can increase the revenue earning capability of the department if the present opportunity is grabbed.

(f) The present policy of the government to take IT to the society for overall economic growth, offers a conducive atmosphere for SOI to provide the basic spatial data to the public domain. The *National Informatics Policy 1999*, of Government of India, considers Information Technology (IT) as an enabling tool for economic development in the country. The policy envisages bringing information technology to all sectors of the society. With this aim it is necessary to identify and remove all forms of impediments on the path of growth of IT based industry, business and services. The issues related to such growth were classified in to four categories

   • Infrastructure and services

- Electronic governance

- Education

- Mass campaign for IT awareness

(g) The NSDI initiative by the Department of Science & Technology (DST) and the Department of Space (DOS) was the first step in the direction of implementation of the above Information Technology (IT) policy. The NSDI initiative stems from the recognition that, while over the years there has been considerable progress in the use of IT in government departments, most of its efforts have been confined to back office computerization. The approach for computerization and IT induction should be from front office application to back office and not the other way. The department can better serve the people through the NSDI gateway by providing the wealth of spatial data of which it is the sole custodian.

Threats

(a) Losing the cliantal base: Spatial data is in great need by the user community for developmental planning. They are also aware that the SOI is having reliable data. They are aware that the it is not able to share the data because of the policies of the government, but are not prepared to accept this situation. If such a situation continues, the users are bound to depend on other data producers for spatial data. Survey of India should realize the gravity of the problem and come out with a solution to meet the user needs, with some practical approach. Otherwise it will be forgotten as the premier producer and custodian of reliable and standardized spatial data.

(b) The good old methods practiced by the Survey over decades need to be modified with the emerging trends to make its products compositive and cost effective

Retrospection

The cartography continues to see greater advances in technology every day. It seems that no matter what discipline one is in, there is no escape from the furthering of use, technology and knowledge. Cartography is not immune to this and in fact it is appreciated because of this reason. Cartography with GIS can deliver the results as per requirement, depending on the use of spatial data even more decisively, depending on the individual users. A map of any shape and size takes on different form for various individuals. Although, one can usually think of a map as being just a static piece of paper, in reality, every map tells a story. NSDI is meant to provide over net the information about the availability of cartographic data produced by various organisations. Perhaps this will prove to be useful for different development purposes. It is a high time to bring cartography closer to the development process at local, regional and national levels.

Survey of India has not been able to keep pace with the ever-changing demands of the market and the public in general due to its rigid governmental bureaucratic structure as well as due to shortage of funds in recent past. Further, it was anticipated that the activities enumerated in the *Prospective Vision* should enable SOI to regain its prime position as the agency for all geospatial data requirements in the near future.

Chapter 2

Transformation

With the change in technology, users' demand, market forces and different applications, the government is forced to have an introspection of its policies and practices about the national mapping agencies. India is not an exception in any way. Transformation of such institutions has become more pertinent because running these institutions have become expensive. There are committees of the parliament, ministries and users who keep on questioning the policies and practices. Similar issues have been raised in other countries as well (Kelmelis, 2002; McGrath, 1982; Rhind, 2000).

The Indian situation is further aggravated because on one hand the age-old system of more than 250 years continues to function; while, on the other hand, there has been a quantum jump in technology, particularly in software and remote sensing technology. As a result, a huge manpower is becoming redundant with the introduction of newer technologies. Field operations, conventional aerial photography, printing, geodesy and data or record management have undergone a sea change. Further, there were several proposals for joining hand with the industry. What should be the course of action under this situation? Should we go for out-sourcing? If yes, which are the possible sectors? What should be the new focus of activities? How to make the maps or the spatial data available to the users in most appropriate way? Accordingly, what should be the transformation model for India? These questions are very intricate, and the possible solutions are furthermore complicated.

In order to address the above issues two decade earlier, one had to understand the then trends and practices of national mapping agency and changing technological scenario as well. The data market had to be assessed. Partnerships with the university and industry were to be evolved. New arrangements had to be made for data dissemination. Manpower was to be trained in different aspects of mapping technology. The old manuals were to be changed which is of course another tedious job. Standard operative procedures were to be implemented. However, the issue was from where to start the transformation process if it was to be done at all.

The *National Informatics Policy 1999*, of Government of India, considered Information Technology (IT) as an enabling tool for economic development in the country. This policy envisages bringing IT to all sectors of the society. Spatial data played important role in effective application of information technology. SOI could not produce updated maps in pace with the development due to the methodology adopted in various stages of spatial data collection and mapping. There were certain other factors which contributed for this state in the organization:

(a) Not been able to maintain the pace with the changing technology.

(b) There was no internal mechanism to regularly review the users' needs.

(c) Accuracy had been the hallmark of SOI, its obsession with accuracy at the cost of the users.

(d) Technological obsolescence due to poor investment in new areas, and

(e) Lack of public relations.

SOI has been known for its accuracy, quality control and strict adherence to standards. Stakeholders had faith in the data produced by the department. However, the organization required to re-engineer itself due to the changing scenario in the application of geospatial data. Then, the user community was gaining experience with the use of digital maps; it became evident that users need different types of products. Further, there was a need that the department responds to the demands of the stakeholders and act as a reliable geospatial data provider, instead of only as a geospatial data producer. The department needed to fulfill the new role. Furthermore, as a government owned national mapping agency, it was to ensure continued availability and to be a provider of high quality updated geospatial data at a reasonable and affordable cost.

Contemporary Status

During 1980s, Survey Training Institute or STI located in Hyderabad was a hub of activities when departmental candidates were undergoing different courses with optimum strength, including 400, 500 and advance courses. The quality of training was certainly of a very high standard, where not only the basic survey techniques were embedded, but also ethics, traditions, honesty and transparency in map-making and cartography were taught. The trainees were explained that the maps then being prepared will be used for future as well. Considering this objective, various survey engineer courses were designed. For example, the *High-Altitude Party* (No.57 Party, North Western Circle, Chandigarh) was designed to be methodical, time conscious, strict disciplinarian, carry out minutest details with highest standards of accuracy. The field surveys consisted of departmental surveys on 1:25,000 scale, rapid verification surveys on 1:50,000 scale and project surveys on large scales *e.g.* 1:5,000, and 1:2,000. Two camps with 7 to 8 plane-tablers were then accepted as a norm. Further for various hydro-electricity projects, two or three detachments under surveyors were always in the field for providing controls tunnel alignment, particularly in mountainous areas.

The above routine continued in the department till 1990s when the operations became affected due to financial crunch. The fieldwork was affected very badly and was almost reduced to half. As a result, whole of the mapping process suffered, since the photogrammetric units lost the incentive of preparing *air surveys sections*, which were not likely to be used for verification surveys in the field for quite some time to come. Meanwhile, there was a ban on the recruitment of Group-D staff as well. Based on the Parameswaran Committee recommendations, their strength had to be kept at the prescribed number. As a result, there were difficulties in carrying out the work as per plan. A sense of helplessness and antipathy started prevailing in the department due to this reason. It is interesting to note that such a large-scale scientific operation had to depend on the Group-D staff for more than 200 years. Nevertheless, the department survived due to its contribution to the security forces and for development purposes. The strength lied in the quality of work both in the field as well as in the drawing offices. However, it was pertinent for the department to ask the following questions to themselves:

(a) Have we been able to sustain and carry forward the legacy?

(b) Have we been able to maintain the quality standards in whatever work we are doing?

(c) Have we been able to live up to the expectations of the society at large?

(d) Have we been able to sustain the pressure of growing requirements of different users?

(e) Have we been able to develop our own human resources who could face the challenges lying ahead?

Transformation Through Training

Survey Training Institute or STI was a single window human resource development solution provider for spatial data generation, interpretation and use in development planning. Its focus was on capacity building by understanding, analyzing, deciding and executing geospatial data. It also tried to educate the trainees how to cope with business environment - the only solution was to change the vision of the national mapping agency and to set their goals in the framework of user satisfaction and to maximize the outputs regardless the profits of their services by limiting them to be cost recovery services. The targets were:

• Customer focused; respond to the rapid changes in the GIS market; offer (just-on-time) diverse products and services that were to be tailored according to market needs.

• To be competitive within the government regulations for the adaptation of 'cost recovery' policy.

• Continue with national responsibilities to help in completing the cadastral mapping.

• Revise the conventional mapping programmes in order to make the completion of nationwide coverage of foundation data, in digital form, as first priority target goals.

• Assess the management of the organization, both at the corporate and operational levels, from business perspective.

- Management by projects and programme.

- Apply modern concepts in operations management to improve performance and quality of services.

The prospective vision included the charter of duties and responsibilities for the vast spectrum of activities undertaken by SOI which demonstrated fully its commitment to be a significant contributor in India's quest for sustainable development as also in providing for the meaningful utilization of its natural resources and creation of measures for preservation of the same. Change was considered to be self-motivational and did not necessarily wait for someone to order for change, anticipate requirements to incorporate change continuously, and anticipate transformation even when things are going right so that it is ready when it comes. One should assess what are the new skills and competencies would be needed. Begin working on them before it becomes necessary, would have a natural advantage. Managing change has a lot to do with our own attitude towards it, for every problem that change represents, there has been an opportunity lurking in disguise somewhere. It is up to SOI to spot it before someone else does. SOI had the responsibility to utilize the potential for making the nation a better place for others who may not have been that fortunate.

Education & Training

Over the years, officers and staff of SOI had pioneered untrodden lands for others to follow and build upon, they surveyed the deepest forests, deserts and swamps, the lowest coastal belt and highest mountains. All these had been possible because of the basic training they had undergone either on job or trained specially for carrying out specialized survey task. Britishers established a Training Centre at Abbotabad (now in Pakistan) which trained the officers and surveyors during pre-independence India. Immediately after independence, a Training Directorate was established at Dehra Dun for the officers and staff. This directorate was subsequently shifted to Hyderabad in 1962. With the advancements in the field of computers, electronic, satellite techniques and remote sensing, the survey technology had also undergone a qualitative change to catch up with modern trend. It was established on 6th May 1967 with the assistance of the United Nations Development Programme (UNDP). Further, the Training Directorate was reorganized as Survey Training Institute (STI) in 1970. STI, apart from catering for internal training needs, it also imparted training for other organizations in India and also for trainees from developing countries. With the introduction of modern sophisticated hardware and software in SOI, staff and members got trained in collaboration with International Institute of Aerospace Survey and Earth Sciences (ITC), the Netherlands. As a result, STI being a premier institute in survey education in Asia had already geared itself to meet the training needs to commensurate with the fast-changing technology not only for the students from within the country but also from neighbouring countries as well.

In a science and technology-based activity, there were several issues to be considered. Some of them were: (a) investment in human resources to enhance productive capabilities; and (b) utilization of those human resources to produce increased and enhanced quality output. Scientists and engineers were expected to be capable of developing and adopting the technology by designing integrated systems, and by understanding complex nature of the spatial data. Furthermore, the educators and trainees were to be capable of transferring the knowledge at various levels in different application areas. However the issue was: What will be the advantages to policymakers, decision-makers and administrators by assessing the political, social, economic and environmental implications using geospatial technology? Nevertheless, STI at Hyderabad was conceived and developed as a premier institution for imparting training in all disciplines of surveying and mapping for trainees from SOI, other sister map-making organizations as well as foreign trainees from neighbouring countries. The institutional set up was to expand its activities considering the following facets:

- It had fulfilled its assigned role for almost 25 years.

- It was proposed to modernize the STI by funding through the departmental resources as well as by entering into collaboration agreements with institutions of repute abroad like ITC, The Netherlands etc.

- SOI had been involved in the research and project implementation in the following disciplines: glaciology, seismicity, global warming, redefinition of geodetic reference framework, sea level studies etc.

- It was involved in setting up test beds and antenna calibration etc. for projectile tracking and remote sensing satellites.

- Also, spatial data structure, transfer formats, GIS software creation, landforms and structured modelling etc.

- Involved in global mapping, national geographical digital data infra-structure, legislations in cadastral surveys.

- Participated in engineering surveys for development schemes, geo-physical and geodetic research.

- Created digital cartographic data bases and their use in GIS and Land Information System (LIS) environment for spatial planning for sustain-able development and environmental plans.

In the then proposed structure of the STI was to be headed by an Additional Surveyor General along with several directors and Deputy Surveyor Generals (4), Deputy Directors (7), Superintendent Surveyors (11), Officer Surveyors (27) and Surveyors (42). Further, following faculties were proposed within this training institute:

(a) Geodesy

(b) Digital Mapping, Cartography & GIS

(c) Photo & Remote Sensing

(d) Topographical Surveys & Land Information System (LIS)

(e) Research & Development

(f) Management

(g) Technical Services & Consultancy

Role of Survey Training Institute

STI imparted survey training at technician, technologist and professional levels. Besides imparting training at basic, mid-carrier and graduate level courses, the institute also conducted advanced post graduate level, management and user-oriented courses in all disciplines of surveying and mapping technology. By then it had more than 20,000 alumni including

600 from other Afro-Asian countries. Table 2.1 indicates nature of trainees during the period 1995-2003.

Year	SOI trainees	Non-SOI trainees	Total Indian trainees	Foreign trainees	Total trainees	Income in Lakh Rupees
1995-96	382	155	537	17	554	NA
1996-97	262	121	383	14	397	NA
1997-98	217	131	348	7	355	NA
1998-99	188	78	266	10	276	NA
1999-20	130	87	217	-	217	34.34
2000-01	111	133	244	18	262	62.16
2001-02	168	296	464	32	496	67.90
2002-03	243	298	541	40	581	55.00

Table 2.1 *Trainees in STI, Hyderabad*

The institute has been headed by an Additional Surveyor General. Earlier it was led by Senior Director and sometimes at the level of Director. There have been other officers at different levels, such as Deputy Surveyor General, Deputy Director, Superintendent Surveyor, Officer Surveyor and Surveyor. Major wings were as follows:

(a) Geodesy

(b) Digital mapping, cartography and GIS

(c) Photo and remote sensing

(d) Topographical surveys and LIS

(e) Research and development

(f) Management

(g) Technical service and consultancy

STI offered courses for complete solution in the field of mapping, provision of control for national connectivity; topographical survey; city survey for planning; engineering survey for road, rail line, tunnel, power transmission line design; cadastral survey by modern methods; photogrammetric solutions: digital and analog; remote sensing input for data collection; and GIS applications and tools. The length of the courses varied from short duration modules to production-oriented modules to basic and refreshers courses for various levels. Further, there was a provision for user oriented extra-department modules as well. However, the regular courses were as follows:

Basic Courses

- Surveying Engineer

- Surveying Supervisor

- Survey Technician

- Cartography Technician

- Managerial Level

    – To assess requirement

    – To set standards and tolerances for any job

    – To plan efficient execution

    – Proper quality control at all levels

- Supervisory level

    – plan assigned task and execute independently

    – quality control as per laid down standards

    – supervision of quality of work by subordinates

- Technician level

        (Data collection, editing and cartography)
    – execute assigned task

Advanced Courses

- In depth knowledge of the technology in different disciplines

- Duration varies from 22 weeks to 32 weeks

- Course fees from Rs 22,000 to 35,000

  - Photogrammetry and remote sensing

  - Advanced geodesy

  - Cartography and GIS

The above mentioned courses included geodesy, reference systems and datums, control surveys, astronomy, global positioning system, electronic total station, modern techniques of surveying, application of GPS, ETS and mobile mapping. The courses related to photogrammetry and remote sensing included analog and digital photogrammetry; satellite imageries and remote sensing; extraction of information from aerial photographs and imageries for updation of maps; cartography and GIS; creation of land information management system; formation of application-oriented GIS; and overlaying of cadastral maps on topographical maps. The faculty were taken from the qualified officers of SOI having experience in related fields including geodesy and astronomy. The courses included out-door practical under real field environment away from Hyderabad.

During the recent past the activities in the institute increased several times due to interaction with users with an option of designing the courses to meet their needs. The structure of the courses had been modular giving more emphasis on digital photogrammetry and GPS. Further, survey related courses were for survey technicians, survey supervisors, survey engineers and the like. In addition, there were short duration courses on:

  - Cadastral control

  - Digital photogrammetry

  - Photo operator

  - GIS applications

- Control survey by GPS

- Office procedure

- Control survey by GPS and total station

- Map reproduction supervisor

- Digitization of cartographic documents

- Digital photogrammetry and remote sensing

- Administration & management

The extra-departmental courses were undertaken for Natural Resources Data Management System (NRDMS, now National Geospatial Programme) of the Department of Science & Technology, Space Applications Centre of the Indian Space Research Organisation, Ministry of Defence, Power Grid Corporation, Andhra Pradesh Government, Karnataka State Remote Sensing Agency and the Government of Sri Lanka. These customized courses included topics like topographical map for sustainable development, coastal zone and disaster management, high resolution imagery for planning, principle of digital map, GIS concept, application of digital elevation model, modern trends in cadastral mapping and datum, coordinate system and projections.

STI provided solutions to the problems related to spatial data capture and their use for development planning. It had been a centre of change, adoption of new technologies and development of geospatial data not only for SOI but also for other institutions in India and abroad. Nevertheless, it was to remodel itself in order to include methods for transformation in surveying and mapping procedures within the organisation. In fact, it was to take a leading role in this direction as well. In the *Workshop-SOI-Towards a Contemporary Renaissance* held in Ramnagar in 2002, following observation was made:

"We have not been able to sustain our growth both in terms of human resources as well as our products. We invite criticisms in most of the forums for our inward-looking attitude and self-defeating policies. We are criticized for our lack of vision, lack of knowledge and stereo typed thinking."

It was expected in this workshop that SOI will assume a new role with NSDI. This initiative was considered as a step towards a quantum jump because it was supposed to bring together the data providing agencies under one roof thereby providing the users with standard data sets. Nevertheless, the role of SOI, despite of this great inventiveness, remained almost the same; and it was felt not to change its basic role, *i.e.*, providing updated topographical maps which could be either hard or soft copy.

MOU Between DST and ITC, The Netherlands

A Memorandum of Understanding (MOU) was signed by the Department of Science & Technology (DST) and the International Institute for Aerospace Survey and Earth Sciences or ITC in July 2001. Earlier, officers from SOI and the National Atlas & Thematic Mapping Organisation (NATMO) were deputed to ITC for various types of courses in geospatial technology. In fact, with the assistance of ITC, STI was established in SOI premises in Hyderabad. This MOU was also meant to strengthen the earlier cooperation between both the organizations. For this purpose, the general areas of interest for both the organizations were the follows:

(a) Interaction on policy aspects of NGDI (later NSDI) at DST (including SOI and NATMO).

(b) Interaction on commercialization at SOI and NATMO.

(c) Institutional support and human resources development, including training the trainers, at SOI and at NATMO in support of NGDI.

(d) Support was required at SOI for producers and users of geospatial data:

  o Specifically education and training to allow the dissemination of geospatial data for applications such as rural development including watershed management, urban infrastructure development and disaster management (specifically monitoring movement of the earth's crust by GPS).

  o Introducing modern technology to implement and sustain NGDI.

(a) Support was thought to be required at the NATMO to introduce modern (Web) technology to disseminate small scale thematic data in support of the NGDI.

o With special attention to the conversion from analogue to digital production processes.

(b) Support the establishing an apex institutional facility of global standards to serve as a Centre of Excellence in the field of GIS and GIS applications within the framework of human resources requirements of the NGDI.

The purpose of this MOU was to provide a mechanism for both the parties (DST and ITC) to work together more effectively in mutually agreed progressive and supportive activities which do not prejudice any relations and agreements with other parties.

Fieldwork and the Organizational Structure

Intensive field operations were essential part the department's work which suffered from fund crunch and shortage of manpower. The status of such operations was as follows:

a) The field surveys were restricted to verification surveys on 1:25,000 scale.

b) Very few or negligible project surveys.

c) Very little compilation on 1:50,000 scale. The result was that 1:50,000 scale maps (which were most widely used) were not compiled or even updated. In most of the cases, the maps were of 1970's vintage and were certainly cause of concern.

d) The extra-departmental work was very less because it was expensive, time consuming and due to delayed responses.

e) Mushrooming of number of private surveying agencies, who could produce, although inferior quality maps were available at cheap rates and also in time.

f) The topographical maps use to give a very unattractive and outdated look while printing technology was going through revolutionary changes.

g) Topographical digital data of whole of the country was available on 1:250,000 scale. In spite of the fact that the department then had digital facilities for more than a decade and claimed to possess state-of-art technology, SOI had digital data of about 350 out nearly 5,000 sheets on the scale of 1:50,000 scale.

The department use to have so-called decentralized system of administration having zones and directorates under them. This arrangement was made keeping in view of quick decision-making and thereby less time taking. However, the results were not as expected. The *zonal system* was not able to meet the requirements of the digital age knocking at the doors. This was due to the reason that powers were not decentralized.

Due to this reason, zones functioned just as out-posts. Each and everything matter was to refer to the Surveyor General's Office or SGO for approval whether it was financial sanction or any other initiative. The directors were also affected in similar way. Directors' conferences became rare over time. Hence, there was practically no forum to discuss the policy, priority and initiatives required for the development of the department.

The induction of the army officers in the department was stopped since 1984. Very few officers were recruited through the Union Public Service Commission (UPSC). As a result, more than 50 per cent of the post at the executive level *Officer in Charge* of a survey party or in short *OC Party* were either vacant or being manned by the Officer Surveyors on *ad hoc* basis or current duty basis. On the other hand, the recruitment was stopped at Group-C level since 1999 as well, and also 50 per cent of the working posts were lying vacant at the level of surveyors, plane-tablers and draftsmen. Then supervisory staff, survey assistants or officer surveyors were not very young. However, some of the departmental officers and staff had undergone basic training in 140,150, 400 and 500 courses. Further, the departmental candidates were promoted from Group-D to Group-C or Division II posts with minimum qualification of class 8th although the basic qualification for Division II staff was raised to minimum 12th class.

New Products and Services

The products had to be updated and were required to meet the emerging needs of the society. After adoption of 1:25,000 as the basic scale of mapping in the Department, the focus on 1:50,000 scale became obscure. Hence, the preparation or updating of maps at this scale got lesser priority. However, the topographical maps at the latter scale maps were much in demand. Nevertheless, there was hardly any difference in the details in the two scales. It was suggested that topographical maps at 1:10,000 scale or larger should be considered instead of 1:25,000, the data of which was mostly required by the users. The planners at the macro level did require maps at larger scales.

With the introduction of computer technology in the country, GIS became a great leveler across the different streams and professions, be it a cartographer, engineer, remote sensing specialist and the like. The products were to be re-oriented in such a way that they fulfil the requirement of GIS applications as well. Further, earlier it was felt that cadastral survey should be included in the charter of duties. Although later is a state subject. Survey of India's strength lied in fieldwork which has also been useful for cadastral surveys. Recently, in *Swamitava Project*, SOI has proved its role in detailed mapping of rural properties.

Generation of digital topographical database was contemplated the biggest need of the time, since it is required by the GIS community throughout the country. The requirement was expected to lay maximum stress on production of data and thereby becoming a very active partner in the NSDI initiative. The *standard operating procedure* (SOP) for digitization required a very deep introspection. There was a requirement of adopting uniform standards in all the directorates and offices.

With the availability of high-resolution satellite imagery, it was being claimed that updating of small-scale maps became a very distinct possibility. In order to update the topographical maps quickly, satellite imagery was considered to be a viable option. Regarding, reproduction of maps, much importance was given to Barco and Map Publishing Systems. It was expected that aesthetically designed publications will enhance the image of the Department. The expected results were yet to come.

Possible Options for Financing

If the data was updated, it was to be more sensitive to the user needs, and SOI was in a position to provide data at a competitive pricing. As a result, the users were inevitably looking for products and the department could have become more self-sustaining. Of course, the SOI was to reconsider its policy about overhead charges in order to be more competitive. The government was also in favour of reducing the prices and generate resources by becoming proactive. Further, *restriction policy* certainly needed a fresh look and be more transparent since the users felt very shy in using the products. *Restriction policy,* it seems, was becoming a disincentive and impediment to sell the products and extend services.

Other Important Issues

It was felt that the SOI should rise to the occasion and reorient and reframe its policies so as to meet the user demands which were completely new in comparison to a paper map. The mission may not change, but the mandate was to change with digital environment. The role of SOI was to be modified keeping in view of the new mapping environment, users' perception and potential application of digital field data.

The digital data was very different from the conventional paper maps. It was in a different format, easily transportable and maneuverable. In this context, some standard policies became essential. NSDI being a national policy and SOI was considered to be one of its constituents, accepting the NSDI policy became mandatory. However, the SOI had to restructure its mandate in view of above policy to meet the demands.

The topographical maps have been the base map for all planners and developers. Thus, necessity of its digital maps become crucial. Hence emphasis was given to complete the digitization of 1:50,000 scale maps covering the whole country became a commitment to the nation. This was an ideal scale for all users for macro planning. However, for micro planning, large scale data from 1:25,000 scale onwards could be provided at the cost of the user. Further, the 1:50,000 scale was selected as a national base because there is not much difference in data between 1:25,000 scale and 1:50,000 scale from users' perspective. However, there may be four times differences in price structure of these maps and user always thought twice in this regard. As a departmental policy, covering the country on 1:25,000 scale was still being perused. However, conversion to digital data on this scale was only to be done at indenter's cost for time being till the SOI completes the digital data base of whole country at 1:50,000 scale. The proposed series was to supply the digital data on NSDI format on WGS-84 datum at 1:50,000 scale.

Conventional map making process was having in-built time-consuming procedures. These techniques cannot be successfully implemented in digital environment. The solution was to adopt soft copy photogrammetry, and satellite stereo model could also be considered to prepare a digital data base. Even though ground verification was must, using photo-verification techniques on latest aerial photographs certainly quickened the updating process and digital maps were produced almost with real time data to the satisfaction of the user. Nevertheless, SOI could not completely do away with old products considering the larger sections of users who were not tech suave and preferred the old product even though change in technology implies modification in form of products. However, use of verification on latest aerial photographs and using latest surveying instruments like Global Positioning System (GPS), total station, Electronic Distance Measurement (EDM) and the like were considered.

Finances required for this purpose could at least partially be generated by undertaking projects. Nevertheless, to achieve this objective, the overhead charges were either be restructured or removed. Such initiatives could have brought down the cost of digital data, except 1:50,000 scale digital data which was priced by Government of India. All other digital data was to be priced at actual cost of survey including reasonable overhead charges. Diversification of activities in order to include cadastral maps, commercial use of assets and disinvestments could be other options as well.

Professional restructuring of SOI staff was necessary to give more emphasis on technical work by supervisory officers who had to deal with both administrative and technical work including drawing and disbursing of funds. It was even suggested to separate the administrative duties and raise separate office directly under Deputy Director (Administration). As a result, there could be only three types of sub-divisions for (a) field operation, (b) photogrammetry and (c) digital data generation. The drawing offices were no more required under digital environment. It was proposed that Directors' post should be of Senior Administrative Grade or SAG. Each printing unit should be provided with digital printing machines to take up digital files for printing.

For capacity building, decentralization and restructuring of staff was felt necessary. To achieve this objective, the recruitment, promotion and transfer was proposed to be within respective zones and responsibilities was proposed to be entrusted to the concerned Additional Surveyor Generals. Further, the Group 'C' designations were projected to be reduced and streamlined. Furthermore, emphasis was to be given on implementation of 1989 rules and recommendations of Vth Pay Commission. Needless to say, job satisfaction and harmonious career progression and belongingness were the hallmark of the progress of the department.

Refocusing the Priorities

SOI had been providing the geospatial data in the form of conventional maps for over 250 years and is the oldest department of the Government of India. It had also been the role model for many such mapping organsiations in the world including the Ordnance Survey of the United Kingdom (Q&C 18). Many methodologies and techniques for surveying and mapping were invented in India and implemented all around the world. Of late, it has been lagging behind in the technology of geospatial information and unable to provide the requisite data in the requisite form for many of the developmental activities. Considering the fact that it has been acknowledged leader in the information technology, this is an incongruous situation which can be rectified by changes in the statute and method of working. It has been a high time to involve the private sector in a big way. Map-making had all along been in the domain of the government.

Reasons for the Present State

SOI had been a monolithic organisation and its branches had themselves grown into full-fledged agencies and departments. While the off shoots have been prospering, SOI has started withering away. There are many reasons for the same:

- Mistaken belief that maps cannot be made from satellite imagery.

- While accuracy had been the hall mark of SOI, its obsession with accuracy was at the cost of the users.

- Lack of direction from the top.

- No perspective plans.

- No accountability.

- Gradual decline in the annual budgets.

- Technological obsolescence due to poor investments in new areas.

- In spite of being the leader in technology having introduced photogrammetry and remote sensing in India, utilization of aerial photography and remotely sensed data was not given priority.

- Disillusioned and demoralized staff including top management.

- Reluctance to induce new management techniques and practices.

- Conflict amongst Group 'A' cadre officers.

- Lack of streamlined recruitment leading to aging workforce.

- Absence of personnel and career management.

- Except for one officer coming on deputation, no outsider for performing varied functions such as finance, marketing, product development, technology management.

- Heads of different establishments conducted the affairs in a personalized manner and did not allow growth of second rung management.

- Lack of recognition to meritorious work led to exodus

- Lack of public relations and often blamed the policy on maps and data.

In spite of all the above, this organisation has taken significant steps in upgrading technology. It was one amongst the first in the developing world to initiate digitization of maps as early as in early 1980's. It took active part in computerization, introduction of GPS and had digital data of the entire country of 1:250,000 scale maps. SOI had good work culture and can could have delivered the results as it had been doing in providing the maps for all major multipurpose projects undertaken by independent India.

An important aspect, generally not appreciated, was that the techniques of surveying and mapping were not taught in any academic institution in India. From group 'A' to group 'D', all staff had to be trained in the department, the skills are thus passed on from one to another in ancient *Gurukul* traditions. It was felt that a time has come when skills were likely to get lost as there was no recruitment for over two decades, and hence the expertise developed over time were becoming extinct. The major advantage of was the ground knowledge. There is no alternative to ground truth. GIS activists were not interested in collecting ground information hence did not valued much about its significance. Moreover, reliable terrain data was required for GIS applications. Further, the benefits of GIS did not be percolated to grass root level unless data of higher resolutions such as 1:1,000 scale maps were available. Thus, SOI had been the savior in geospatial data acquisition ranging from 1:50,000 or 1:1,000.

> It was also felt that SOI should consider ways and means to provide data on 1:1,000 scale for all urban and semi-urban areas as well as all villages. For highly congested areas scale of 1:500 was considered as well. Till security issues were resolved, the red plate and black plate could have been digitized which appears to have a huge demand for many applications such as location-based services. Further, it was also observed that the industry has matured over a period of time; and SOI should resort to outsourcing of some of its activities; and at the same time, in house capacity building should also be strengthened to meet societal needs. Immediate investments were proposed to be made in newer technologies such as ALTM, soft copy photogrammetry and the like.

Action Plan

With a view to bring radical changes in SOI and to bring down the strength to a level below 7,000 employees from the existing 16,000 and to increase private sector participation in all activities of surveying and mapping including quality control, the following steps were suggested:

1. Establishment of State GDC:

- Immediately do away with party concept. As such there were not enough personnel in most of the survey parties.

- To locate state GDC in all state capitals. Most of the capitals have SOI presence which can be converted to state GDCs.

- In a few states such as Tamil Nadu centre was raised. In some places such as Maharashtra, the existing set up in Pune was converted to GDC as it was not possible to find place in Mumbai.

- A detailed state wise plan was worked out.

- This activity was to be completed by 31 Dec 2002.

2. Capacity Building:

- Action to be taken to immediately recruit personnel for groups 'A', 'B' and 'C'.

- For Group 'A', a recruitment was to be organized to recruit at least 100 officers: 50 for civil cadre and 50 from the army.

- This activity should also be completed by 31 Dec 2002 so that training could commence by 01 JAN 2003.

- A thoroughly revised and shortened training were to be organized and the new training programmes design was to be completed by 31 Dec 2003.

- There was a shortage of trainers which could be temporarily met by outsourcing or hiring retired employees of SOI.

- In additional training could be organized at various state GDCs for Group 'C'.

3. Outsourcing Survey Operations:

- All survey operations right from data collection to printing could be outsourced since the aim was to bring up the currency of the data.

- A template was to be designed for all such outsourcing activities and streamlined procedures were made to bring transparency and efficiency in all such activities.

- If necessary, professional help was proposed to be sought from designing competitive bidding processes which were totally transparent.

- Traditionally. SOI has been shy of outsourcing, which could have shaken off with transparent procedures in place.

4. New products and services:

- All along the users' needs were never taken into consideration except the requirement of the armed forces.

- The maps designed for the army were also being used for civilian developmental needs.

- This also led to non-availability maps as they contained information which were not required, but sensitive.

- It was realized that in the digital domain the required data can be given to the civilian user wants without compromising security.

- Some of the products and services that could immediately be started are given below:

• Digital data of 'Where is it' could be sold to many companies selling software such as horoscopes, where latitude and longitude of places is required. Every copy of software sold should fetch some royalty.

• Road map of India with some additions to the marketing software packages for purposes such as traveling salesman, distribution of goods and services.

- Providing black and red details on 1:50,000 or 1:25,000 scale topographical maps for location-based services.,

- All town maps and guide-maps in digital from after updating could be sold to utilities such as cellular phones, domestic gas distribution, location of services (post box, telephone booth).

- All boundaries up to village (wherever available), could be packaged and sold to number of solution providers who advice companies and offer solution in their marketing strategies. This data would also help state and central government departments such as Registrar General of India (Census) in e-planning and e-governance.

- STI could market itself offering courses in various multi-disciplinary technologies. It should get transformed into *deemed university* so that the degrees and diplomas issued could fetch jobs.

- SOI possessed rich knowledge base in digital domain as well. Many private sector companies are engaged in digitization, digital photogrammetry etc. could be provided consultancy and advice. SOI should encourage growth of private sector instead of picking holes in their work.

- It may be unexpected to note that SOI has spare printing potential in spite of the fact that maps fair drawn a decade ago are yet to be printed. This potential could be used for commercial purpose.

- SOI should be proactive in securing outside work. However, credibility has to be demonstrated quickly so that others can trust that the delivery is on time.

5. Personnel Policy and Administration:

– SOI does not have a personnel department to look after staff management.

– Such a nucleus should be created by 31 Dec 2002.

6. Technological upgradation:

– SOI had all along been the leader in technology. Even when computers were being introduced, SOI was amongst the first few to start computerization. Even orgnaisations such as Computer Society of India came into the being mainly due to active involvement of SOI (Fig 2.1).

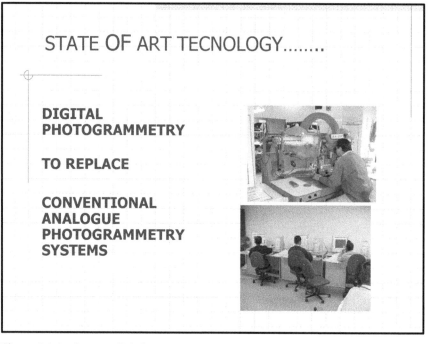

Figure 2.1 *Analogue to digital*

• SOI was the one amongst the first to start creation of digital geospatial data. In the recent past, there were slippages and the days of fast changing technology, it does not take much time to go in technological obsolescence.

• There was a need to invest at least Rs 200 crores in the next five years to invest in new technologies such as softcopy photogrammetry, air borne laser terrain (ALTM) mapping, image processing for mapping and the like.

• It should be a target to get every person in SOI a computer savvy even if he is otherwise sound.

The above observations were considered and found to be pragmatic; and it was noticed that SOI should join the then evolving superhighway of geospatial information as a part of information technology or otherwise. This objective cannot be achieved overnight. However, it can be accomplished by having a two-pronged approach *viz.,* (a) short term, and (b) long term. Further, the staff strengths and capacity building by streamlined training programmes for the next 10 years need to be prepared. If required, the job of drafting long term solutions could be left to professionals who would usher in the changes and managing them.

Long Term Goals

One of the long-term objectives was to make available geospatial data of resolution 1:1,000 in digital form in real-time to the users. Cost recovery, if any, was to be done in an automated process using credit card or smart card or debit card. Making copies in magnetic media and dispatching them should also be part of the automated process. The long-term goal was also to convert Survey of India completely digital by 2005.

Short Term Goals

- By 10 April 2003, all maps on scales 1:50,000 and 1:25,000 were to be digitized.

- SOI node for NSDI should be on the net giving the availability of data and metadata by 15 Aug 2002.

- Establishing GDCs in all states by 31 Dec 2002.

- Introduce new products and services by 01 Jan 2003.

- Recruitment of at least 100 officers at Group 'A' and 500 at 'B' and 'C' level by 01 Jan 2003.

- Induction of new technologies and establishment of supremacy in geospatial data acquisition, management and dissemination.

Proposed Vision for Survey of India

Keeping in view of the issued discussed above, an initial attempt was made to propose a vision, mission and objectives for Survey of India. However, as we can gather, after a series of discussion, a lot of modifications were made. The initial attempt was as follows:

Vision

SOI's vision is to be the cardinal apparatus for rendering services and principal provider of geospatial data to the user community in the country.

Mission

SOI is dedicated to the advancement of theory, practice, and application of geospatial data, and promotes an active exchange of information, ideas, and technology innovation amongst the data producers and users who will be having access to the geospatial data of highest possible resolution (say 1:1,000) at an affordable cost in the real-time environment.

Objectives

- SOI should be vibrant, that is responsive to the needs of user community in India.

- To promote professionalism, incorporating high standards of ethics and conduct among geospatial data users.

- To provide a forum for all the professionals to share knowledge and experience.

- To stimulate, encourage and participate in research, development and application of geospatial data.

- To be active in all the areas of application of geospatial data through special interest groups which would be dynamic to accommodate the emerging technologies and applications.

– To contribute to the formulation and organizing of the education and training programmes needed by professionals, users and the public.

– To actively participate and advise the public bodies including the government in the formulation of policies affecting the professionals.

Requirements from the Industry

The requirements from the Indian geospatial industry were estimated based on series of discussions. The outcome of such initiatives was as follows:

– To become front agency for bargaining projects: national and international

– To provide fundamental data sets in digital form for value addition

– To jointly bid for projects etc.

– To bundle SOI data with software

– To certify data quality

From the above details, it may be gathered that the expectation from the industry was much beyond of the conventional activities of SOI. On the contrary, Survey was concentrating on (a) staff reduction; (b) higher financial provision for modernization; (c) commercialization; (d) outsourcing; and (e) revenue generation. The financial scenario was as follows:

– About 90 per cent of the annual expenditure goes on non-plan sector out of which maximum is on "Salary" head

– Revenue generated every year was about 20 per cent.

– Uneconomic small units all over – consolidation was required.

– Reasonable sum was paid for rents and taxes for hired buildings.

– Large estates were underutilized (Appendix V)

– Surplus staff in some sectors

In order to achieve some of the above-mentioned objectives, it was considered necessary to introduction of new technologies, take up innovative products, new type of partnerships with industry and academia' and finally a new vision. The idea was that SOI dedicates itself to the advancement of theory, practice, collection and applications of geospatial data, and promotes an active exchange of information, ideas, and technological innovations amongst the data producers and users who would get access to such data of highest possible resolution at an affordable cost in the near real-time environment.

Expectations

The characteristics of the geospatial data set were changing. First and foremost, in order to meet users demands effectively, the capacity for the real-time collection, synthesis and access must exist; data currency was essential. The data should be scale less, seamless, without artificial boundaries, and linked to a time component that became critical to many applications, for example, traffic flow management, routing and delivery, and tidal and marine traffic. Moreover, as technologies become more advanced, geospatial information was considered to be both more readily available and in greater demand.

There was also a growing trend toward the collection and integration of non-traditional data using secondary reference systems like voting, culture and housing patterns, gender, sales and industry. Furthermore, as technologies and applications became more globally used, geospatial data was to spread to and originate from non-traditional sources such as the voluntary sector, health councils, communities and peoples. However, regardless of what data were collected by whom, unless they were easily and readily accessible, their value diminishes. Hence, the importance of an exceptional geospatial data infrastructure was felt. Furthermore, a well-developed national information infrastructure, enabling the dissemination and sharing of valuable, geographically referenced information, and with an ever-increasing audience of businesses, entrepreneurs, students and researchers, and communities, was widely accepted as an essential asset for any country to maintain and to advance its social and economic well-being. As such, geospatial data and the infrastructure in which were supposed to be organized, was considered to be an expertise in its own right within the rubric of this technology.

The geospatial data was expected to play an important role in resource and land management. Geospatial products like maps, atlases, charts, remote sensing data; GIS were being used frequently to indicate that how and where the possibility of land development was needed and how various earth resources located in an area and how these were spatially interrelated. Whereas GIS could help in managing and making developmental plans for effective management system. Images, GPS and GIS were considered to be extremely important technologies for addressing solutions of various users. The technology of surveying and mapping is knowledge-based and having no bounds or limits even at that time and was having the possibility to evolve in numerous directions. The technology advancement was to determine the direction and strength of this evolution over the next few years. As such, it was very difficult to describe how the technology will shape at any point of time and degree of certainty.

New Vision & Missions

After having a series of discussions, it was found that there is a need for change. The focus was to shift considering the technology, market and government policies. As a result, a new vision was worked out (Hadley & Hammond, 2003; Pande, 2003). The new mission of this national mapping agency was:

> The Survey of India will take a leadership role in providing customer focused, cost effective and timely geospatial data, information and intelligence for meeting the needs of defense, sustainable national development and new information markets.

The mission obviously indicated about the paramount role SOI has to take in providing geospatial data. This implies that SOI was to become a leader in this field not necessarily by generating the data. Hence the new focus pertained to becoming a pivotal element in generating the data by the associated agencies as well. A corollary to this was to develop partnership with industry and universities. Further, the focus in a way had also shifted from the conventional printed map to digital data, which was to be processed for various applications. In order to understand the emerging scenario in the country, a separate type of specialization was required within the organisation which was somewhat different from topographical mapping. Hence, a window for this purpose was required as well. Officers were to be trained. Furthermore, defense continued to get due priority, but two new types of users have been identified as well: one for sustainable development; and the second, for information markets. Sustainable development was considered to be closer to the issues related to resources and the environment. This aspect had been in the forefront of the remote sensing institutions in the country which got a boost with successful launching of the Indian remote sensing satellites. Under the aegis of the Department of Space, several national and state level remote sensing application centres have been dealing with such issues. The activities of such institutions were studied in order to have a sharper focus for SOI. In the same way, the new information markets were either linked with the industry or infrastructure, including cadastral mapping. The industry was heavily engaged with the mapping related to infrastructure while cadastral mapping continued to be a state subject under the federal structure of the governance. What leadership role SOI could have taken in this field was still to be worked out with the concerned agencies. However, these issues continue to be discussed at length even today.

Aims

Keeping in view of the above vision and missions, the new aims of the institution was worked out which were as follows:

(a) SOI will be vibrant, *i.e.* responsive to the needs of user community in India.

(b) To promote professionalism, incorporating high standards of ethics and conduct among geospatial data producers.

(c) To provide a forum for all the professionals to share knowledge and expertise.

(d) To stimulate, encourage and participate in research, development and application of geospatial data.

(e) To be active in all the areas of applications of geospatial data through special interest groups which will be dynamic to accommodate the emerging technologies and applications.

(f) To contribute to the formulation and organizing educational and training programmes needed by professionals, users and the public.

(g) To actively participate and advise the public bodies including the government in the formulation of policies affecting the professions of surveying and mapping.

It is not that the SOI was addressing these issues for the first time. However, in the scenario where defence requirements were given priority, the understanding of the needs of the user community got a back seat. Further, there were two specific types of requirements: (a) projects surveys, and (b) general public. Some attempts were made for example to produce visibility series, tourism series, town maps, trekking maps, district planning maps and the like. In some cases, seminars and discussions were conducted to understand users' requirements.

The aim of promoting professionalism and ethics among geospatial data producers was not an easy task and same issues more-so-ever still continues. There were agencies under the government which had their own norms developed over the years. To get them changed, lead to a lot of discussions and insight. In addition, there were several industrial groups who were the new players in this field. They were engaged in market driven activities. Their activities were technology driven as well. In order to bring some semblance, the National Spatial Data Infrastructure was established. There were special committees within NSDI appointed for different purposes. Nevertheless, the role of SOI in monitoring such activities was to be decided. The same set up was also meant to provide a forum for all professionals to share knowledge and expertise in the field of geospatial information.

Till then the objective had been to prepare accurate maps and later spatial data. The participation in research and development was limited in this sphere of activity. But encouragement, promotion and applications of geospatial data required more openness, more interactions and more understanding of research requirements. In order to do so, a new approach became essential in the functioning of the organization. User conferences, participation in the seminars, publishing status papers and reports and willingness to change became most essential.

Considering different various applications, the formation of special interest groups in the traditional organization like SOI was a new approach. Application oriented mapping has never been a priority. But it was strongly felt that user requirements have to be understood and the policy and contents have to be changed accordingly. Data had to be user, application, technology and market driven. The training programmes for such transformation in mapping and as well as training for users turned out to be most important and integral part of the change process. In any case, SOI had been associated with different academic and professional bodies for the promotion of mapping technology. In fact, such involvement had sometimes been more than desirable. In view of the vision, missions and aims for a national mapping agency, an attempt was made to identify the areas of concern considering needs of user community.

Transformed Goals

In an age-old institution like SOI, transformation process has to be initiated in all fronts (Krishna, 2003). New products, business opportunities, training, technology, collaborations and organizational were to be considered in combination. The emerging scenario was complex but pertinent for transformation. The new options were to be accepted as opportunity – not as a threat.

The topographical maps on different scales were not meeting the requirements of planning and infrastructure development in rural and urban areas. Therefore, it was planned that SOI would provide digital spatial data on scale commensurate with the requirements of planning for such areas. The methodology adopted in various stages of mapping was in any case outdated. Latest methods of data capture and processing using digital techniques was then the required. In this context, all maps on scale 1:50k and 1:25k scale available on paper form, which were cumbersome to use, need to be converted into digital form to provide framework spatial data to customers in NSDI platform. Topographical maps needed to be updated using inputs like high resolution imaging data or rapid ground truth checking with digital techniques on a fast-track mode. It was then targeted to transform SOI completely digital by 2005.

The data collected for ground truthing was to be done by field methods but using the latest available techniques like GPS, total station, digital level etc. In order to speed up the office data collection, the available inputs like aerial photo, satellite imagery, Airborne Laser Terrain Mapping (ALTM) data or other acceptable sources were considered to be used. Large scale aerial photographs on 1:5,000 scale was required for mapping urban areas; and 1:15,000 scale for rural areas which was to facilitate generating digital data on larger scale with required accuracies. ALTM technique was being tested to generate geospatial data for large-scale digital mapping. SOI was also planning to use mobile mapping techniques (palmtop with GPS) for data collection instead of conventional plane table methods, so that the data updating and generating digital data can be completed without much loss of time. Updating of existing 1:50,000 or 1:25,000 scale maps was a huge task for the department. There were about 5,000 maps on 1:50,000 scale covering the country. Further, there was a plan to update all these maps using high-resolution imagery like IKONOS, SPOT etc.

The present method of going through various stages during the map reproduction in bringing out the final hard copy map required a thorough review. The method of reproduction was changing from existing method of plate making to use pre-sensitized plates, which worked out to be cheaper and time saving. Proof examination in reproduction office was to be dispensed with. Units responsible for data processing were to complete the whole process and then submit the material for final printing in the presses. This was to reduce the movement of material between printing and processing offices. SOI then possessed four dedicated printing presses located at Dehra Dun, Kolkata, New Delhi and Hyderabad.

The department had set target to provide spatial digital data for existing 1:50,000 scale topographical maps and make it available to stakeholders. It was also committed to provide precise control points at closer spacing using GPS duly connected with leveling datum and monumentation. Further, the department planned to densify Ground Control Points (GCPs) using GPS and other means for large scale mapping for the areas covered by cities or towns on various scales larger than 1:5,000 within reasonable time. Initially, the GCPs was to be provided at a spacing of 10 km and further densify for rural and urban areas on requirement basis.

Till the end of 2004, about 3,500 geodetic horizontal control points and about 4,000 plus precision and high precision height control points were established in the country. These control points were although sufficient for extending topographical points for mapping on scale 1:50,000, but considering the requirement of large-scale mapping, there was a need to densify the precise control for getting quality digital data. SOI planned to use the existing control as reference and densify it by using GPS techniques. Such densified control was to be provided in a fashion to have about 120 points in each map sheet area of 1:250,000 (110 km to 110 km) within the next two years' time for entire country.

It was then proposed that SOI to equip itself with GPS, digital levels and total stations immediately and start densifying the control work. It was in the process of acquiring state of the art digital data capture and processing equipment and complete the generation of updated digital database on 1:50,000 scale by 2005. ALTM technology was proposed to acquire and acquisition of high-resolution digital data for urban areas. In the process, the department switched over from high volume production of hard copy maps to provide these maps on demand or in soft copy form as per users' requirements. For this purpose, the state level directorates were established to take care of mapping requirements of the sates. Survey Training Institute was geared up for capacity building.

SOI developed state level SDI for generating high-resolution digital data in collaboration with state governments. It was proposed that it will actively establish expertise in public and private domains to eventually enter into strategic alliances. SOI node of NSDI was then in the process of coming to the existence shortly.

SOI organized education and training programmes needed by the children, youth, professionals, and users and for the public. STI in Hyderabad was in the process of identifying the areas of skills required in view of IT policy of the country. It evolved new training programmes to cater needs of such skill development in the field of geospatial informatics in order to improve effectiveness of the system. The training of personnel in latest digital techniques was taken up. STI was also geared up to organize on job-need-based training to develop skills for acquiring geospatial data and its processing. Workshops and courses of suitable duration were to be organized to facilitate participation by various sections of the society. Indian universities were approached for recognizing its longer duration training courses.

Regarding business strategy, SOI had been lacking even two decades ago (Nag, 2002). The guiding principles was to take account of commercial pressure, continuously improve performance, and remain customer focused. In order to achieve this, SOI required to increase contacts with customers through partnership. The collaboration was to be on partnership mode or contractual mode, both within the country and abroad. There was supposed to be a strategy of regular monitoring of geospatial market, which could be done through a professional agency to assess potential market share of SOI. Hence, it employed an agency to improve its marketing capability as well as to understand the potentials.

Several methods were being worked out for profit sharing, cost recovery, and on royalty basis. These initiatives were centered on different models including that of the Ordnance Survey. New products were being identified and partnerships had been worked out (OS, 1999, 2001 & 2003). In fact, consultants were to be appointed soon who could help in identifying new markets, products and services.

Conclusions

Geospatial data became more and more critical for effective planning and decision-making. With the availability of modern technology, it was possible to exploit technology to good advantage, obtaining new products and services faster and cheap as compared to conventional techniques. There were several journals then coming out which were implicitly or explicitly pressing the national mapping agencies to provide the spatial data easily. Further, to switch over from ages long methods and tradition of the department, obviously there was some reluctance from various corners to adopt new techniques due to their expertise in conventional fields. However, digitizing maps in computer environment was soon mooted in all the directorates of the Survey. The department adopted latest state of the art technology in data collection, processing and management.

As the national mapping agency of the country, it was then aimed to provide geospatial solutions to all user community timely and in cost effective manner. At the same time, it was recognized that there was no monopoly in good ideas and practices. It was expected that India will be a role-model for other national mapping agencies in the world considering transformation in the methodology, techniques and organization. This has been an ongoing process and will continue.

Chapter 3

Reorganization & Restructuring

In the era of globalization and liberalization, there is an urgent need to restructure SOI in order to make it a viable, compact and vibrant organization by reducing layers of commands and downsizing the workforce and the staff. Of late, a number of companies have entered in the field of surveying and mapping. Several of these companies are managed and manned by retired SOI officials. Many governmental and non-governmental organisations (NGOs) are getting aid and assistance from financial institutions. Surveyed map or plan had been a perquisite for such getting such aids. SOI had not been able to meet such demands owing to lack of will and manpower. It had the scope to accept a leading role in this direction. On the other hand, it could also tap the technically qualified manpower available in the market at cheaper rate. However, considering the availability of manpower and the status of technology, what should be the size of workforce to discharge the charter of duties assigned to SOI? Similar questions have been repeatedly cropping up from time to time. Hence, the issues related to reorganization and restructuring have been always high on its agenda. Further, the shifting of goalposts has made matters more complicated.

Scope for Restructuring

The restructuring exercises need to have a prospective plan of the department for at least next ten years. The restructuring of the department was required to be done keeping in view the vision, missions, prospective, goals and priorities. It became imperative to merge certain cadres, unwanted posts to be declared surplus, and at the same time create some new posts. The capacity building in the organisation included following measures:

(a) Immediate recruitment at all levels. It takes 5-6 years for a survey professional to be really effective.

(b) Most of the officers manning senior level positions require wide exposure and therefore they were supposed to be given adequate opportunities.

Over the years, several proposals have been submitted to the controlling ministry for reorganization and restructuring. Keeping in view of the then situation of manpower, requirement of better administrative efficiency, technology and nature of demand of geospatial products, these proposals were formulated. Some of the demands were accepted by the government such as abolishing of zonal structure, opening of geospatial data centres (GDC) in each state and separation of printing related activities from the GDCs. At the centre of the whole setup has been the Surveyor General's Office. Traditionally, this seat of administration had been decorated with several additional and deputy surveyor generals and other senior officers with specific functions and responsibilities. The then set up in the Surveyor General's Office or SGO was as follows (Fig 3.1).

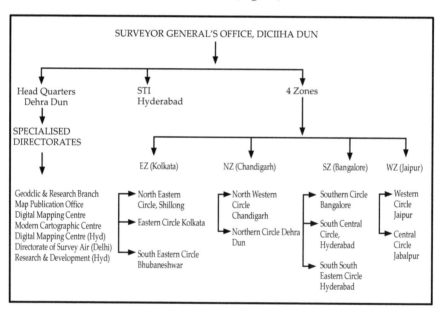

Figure 3.1 *Organisation of Survey of India in 2002*

The location of different SOI offices has been depicted in Fig 3.2. Some of the smaller offices no longer exist while some have been upgraded to state level geospatial data centres (GDCs). Offices located in Ambala, Varanasi, Palghat and the like have been merged with the nearby GDCs.

Figure 3.2 *Location of Survey of India offices*

Status of Survey of India in 2001

It would be a matter of interest to know what the status of the SOI at the turn of millennium was. The restructuring and re-organisation exercise can only be done in a better way if the information about the traditions, procedures, fieldwork methods, application of technologies and the like were comprehended. The main features were as follows:

- Technical work was based on age-old traditions and manuals (*vide* Preface).

- Directorates and lower formations were having depleted strength as per the earlier structure.

- Fieldworks were based on old methods with tents, camps and *khalasis.*

- Application of remote sensing and digital photogrammetry was yet to be introduced.

- Digital techniques were limited to two digital centres and Modern Computer Centre (MCC).

- Digitization of 1:250,000 scale was only completed (394 sheets).

The Surveyor General's Office (SGO) was only looking after administration, and technical work was left to the Directors located in different states. The technical work was organized through zones, circles and directorates with practically no accountability. There was no fixed updating cycle for the topographical sheets. As a result, the out turn was very low because of practice of conventional methods. Maps were decades old. Modernization was in limited places only. The total set up was as follows:

- Zones 4

- Directorates 18

- Specialized directorates 7

- Units119

   - Field units (Topographical parties) 58

   - Photo unites 17

   - Drawing units 17

   - Specialized units 19

   - (G&RB, DSA, STI etc,)

   - Printing groups 5

   - Workshops 3

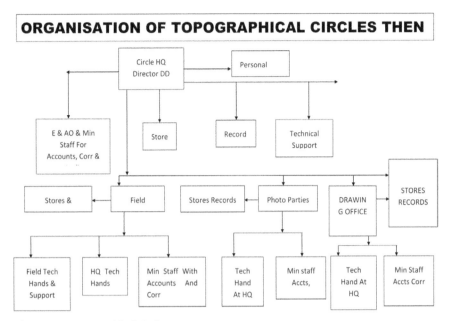

Figure 3.3 *Topographical circles*

The conventional structure of a topographical circle has been depicted in Fig 3.3. It used to have establishment and accounts office, stores, records, photo parties, drawing office, and units for field and technical support. Regarding the manpower, a lot of consideration was necessary. The relationship between army and civil officers was not always harmonious. Associations were anxious with their welfare but critical about the functioning of the organisation. There were delays in conducting the departmental promotion committee (DPC) meetings which created problems for the staff in their career progression. As a result, lots of legal cases piled up. All these issues and their combination created hindrance in the functioning of the organisation. Due to these reasons, the technical administration of SOI was not that effectual. The development issues could not attract much attention, and the establishment was not exactly sensitive to public requirements and market forces. Hence, restructuring and reorganization became essential. Some of the following steps were taken to circumvent these problems:

- Zones, circles and parties were abolished.

- Technology based fieldwork techniques implemented.

- Training programmes for lesser duration were evolved.

- Digitization work was initiated in full swing.

- NIIT/Aptech students were engaged.

- In SGO, a Cell was created to demonstrate the application of modern technologies.

- Remote sensing and digital photogrammetry techniques were introduced.

- G&RB was geared up for densification of control points which was essential for large scale mapping.

- New manuals and standard operating procedures (SOPs) based on digital techniques were introduced.

- On job training programmes were initiated.

- DPCs were conducted to reduce the backlog.

- Consultants were engaged regarding restructuring and commercialization.

Why Reorganisation?

There have been two major streams among the Group A Officers in the organization: (a) Army, and (b) Civil. The induction of army officers was stopped for last 18 years due to several reasons. The civilization officers were recruited through the Indian Engineering Services. The recruitments at lower levels, which were the main work force, affected the production adversely. The scenario for the period from 2001 to 2005 has been depicted in Table 3.1.

Cadre/Group	A	B	C	D
Sanctioned posts	315	627	7,418	7,917
2001	219	475	5,992	5,910
2002	196	421	5,616	5,634
2003	172	365	5,262	5,274
2004	159	310	4,802	4,861
2005	151	283	4,206	4,474

Table 3.1 *Depletion of manpower, 2001-05*

From Table 3.1 it is apparent that the sanctioned posts were having four broad categories of cadres. It was heavily tilted towards Group C and Group D posts who were more required for labour intensive fieldwork methods. The scientific and technical cadres were under the Group A and Group B. Due to different reasons, their numbers were also getting reduced. On the other hand, new technologies were knocking the doors which required qualified and trained officers and staff. Nevertheless, the reorganisation exercise could not be performed in isolation. It had to take into account of the shortcoming of the existing system, cadre problems in different categories which were more than eighty, and interaction with the users. In Fig 3.4, some of the issues have been highlighted.

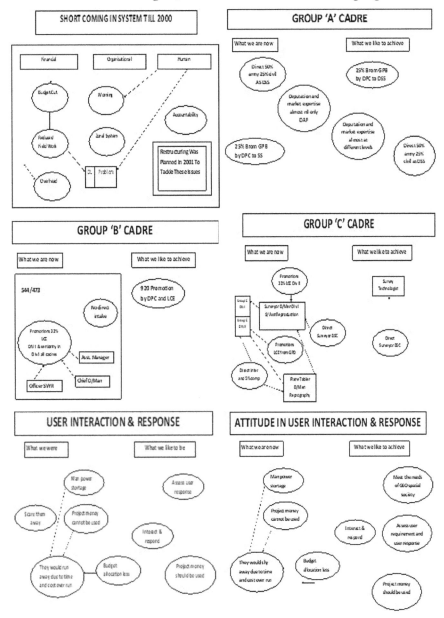

Figure 3.4 *Status and expectations*

Considering different sectors of the scientific and technical work, an estimation was made about the manpower available in about 2002 and manpower required to manage the assigned activities. They included densification of control points, digital photogrammetry, fieldwork for collection of topographical information, supervision, transformation of data into digital form, and management of data and related actions. Table 3.2 gives an idea about the situation then prevailing in SOI. The required manpower was 21,037 – mostly required for field data collection. Densification for geodetic control work also required substantial numbers. However, then only 2,899 persons were available. It was expected that after restructuring this number would increase to 5,354. Obviously, the changing technological scenario, modifications in field operations and new requirement of data transformation were taken into account.

Work	Required Manpower (Per Year)	Available (Per Year) At Present	Availability After Restructuring (Per Year)
Densification of Control	5816	337	2400
Digital Photogrammetry	1,258	322	400
Data Collection	16,100	810	810
Supervision of Work	1,414	474	788
Transformation of Data to GIS	1,372	900	900
Management	77	56	56
Total	21,037	2899	5354

Table 3.2 *Manpower required and availability*

Initial Attempts for Reorganisation

Two consultants were engaged to suggest possible restructuring and reorganisation of this age-old organisation. Dr I.P. Gabba recommended career advancement at all levels by matching and saving in the budget. However, it did not take into account the newly created state level geospatial data centres. Nevertheless, some recommendations of this report were taken into account for restructuring the Survey of India (Table 3.3).

S. No.	Designation	Pay scale	Number of posts		Mode of filling
			Civilian	Service	
1	Surveyor General of India SGOI	Rs 24,050-26,000	1	-	By Government
2	Sci 'H'/ Addl. SGOI	Rs 22,400-24,500	3	-	Vacancy based promotion of combined civilian/service officers based on merit failing which by direct recruitment
3	Sci 'G'/ Director	Rs 18,400-22,400	13	6	Vacancy based promotion of civilian/ service officers against posts mentioned in column 4 based on merit failing which by direct recruitment
4	Sci 'F'/ Addl. Director	Rs 16,400-20,000	20	20	Ditto
5	Sci 'E'/ DD	Rs 14,300-18,300	30	30	Ditto
6	Sci 'D'/ Addl. SS	Rs 12,000-16,500	40	40	Ditto
7	Sci 'C'/ SS	Rs 10,000-15,200	51	54	Ditto
8	Sci 'B'/ DSS	Rs 8,000/13,500	144	-	By recruitment/ promotion
	TOTAL		302	150	

Table 3.3 *Revised cadre structure by I.P. Baba*

Note: Posts at recruitment level (Rs 8,000-13,500) be filled progressively over 5 years period

The above number of 150 service officers included 97 posts deployed in army. Further, it was suggested that mixing of the cadres creates unnecessary problems especially when both cadres are governed by different set of perks, disciplinary and conduct rules. Hence, it was recommended to have separate cadres both for civilians and service (army) officers. The civilian officers included isolated posts were proposed to be brought under *Indian Survey Service* as well.

The second report was by M/S Feedback Strategies. They were appointed to assess the market requirement and potential of geospatial data users. Some aspects of organizational restructuring were touched upon though it was not exhaustive. Here also some of the suggestions were considered in such exercise. In addition to the above initiatives, a Steering Committee was formed under the Surveyor General with following task forces:

a) Technology,

b) Market,

c) Systems and processes,

d) Organizational transformation.

Reorganization Proposals

Apart from the external agencies, several proposals were made not only by the survey administration but also by different unions and staff associations. Obviously, the concerned officers and staff had their issues in mind while doing so. Moreover, the Department of Science & Technology also encouraged to have the views of all concerned. A number of seminars and meetings were held for this purpose. It was a difficult proposition to have a holistic view of the situation satisfying all the stakeholders. Some of the proposals have been included here.

Reorganization Proposal – 1

Group A Cadre

The mode of recruitment and distribution of various posts in Group 'A' cadre was included in the existing Recruitment Rules of 1989. They were to be followed in toto. In order to improve efficiency of the department and to attract qualified manpower, the structure of Group 'A' cadre was proposed to be restructured at various levels. Cadre restructuring of SOI Group 'A' CCS (Central Civil Services) was suggested after taking into consideration the crucial percentages of posts at various levels in organized Engineering Group 'A' CCS & AIS (All India Services). The distribution of posts at various levels in Engineering and other organized Group 'A' Central Civil Services & All India Service (CCS & AIS) officers was also suggested.

Surveyor General of India

As per the recommendations of the Fifth Central Pay Commission, the Head of the Department of an organized services had to be placed in the rank of Special Secretary to the Government of India, in the pay scale of Rs.26,000/- fixed. Necessary action by cadre controlling authority in this regard was to be taken while restructuring SOI. All administrative and financial powers were vested at this level may be provided to Surveyor General of India (SGI or SG) for discharging his duties as head of the department. At the headquarters, SGI was to be assisted by two Surveyor Generals (SGs) in the pay scale of Rs.22,000-600-26,000. SG (S&M) was proposed to exercise administrative and technical control over ten Regional Additional Surveyor Generals. SG (TD) was to exercise administrative and technical control over remaining eight Additional Surveyor Generals.

Additional Surveyor General

All the Regional Additional Surveyor Generals (ASG) were to have a technical, administrative and financial control over directors placed under their control. The financial powers of head of the department were to be transferred to all regional heads as applicable in other organized services. They were to be *Appointing Authorities* for Group 'C' cadre in that region. The procurement, outsourcing and consultancy pertaining to that region was to be controlled by them. All the officers of the Group 'B' and Group 'C' level were to be transferred by them within their region as per the requirements of work. About 40 sheets of 1 x 1 area was to be allotted to each Additional Surveyor General (ASG).

Director

Directors was to exercise administrative and technical control over the Superintendent Surveyors (SSs) placed as officers-in-charge of offices and they were to function from the office of the regional Additional Surveyor General. Wherever necessary, they could have performed from other places as well. In that case the unit at that station was to draw their pay and allowances separately.

It was felt that the number of trades at Group 'C' level must be reduced to a bare minimum. A few suggestions were as follows:

(a) All the trades of Division I & II staff of Group 'C' such as Survey Assistant, Plane Tabler, Air Survey Draftsman, Topo Auxiliary, Trig, Computer, Record Keeper, Store Keeper etc., be merged into a single trade of "Topographer".

(b) "Cartographer" can be another cadre in Group 'C'. All the draftsmen in Division I & II can be part of Cartographer trade. Total strength of Topographer' and 'Cartographer' may be kept about 2,000 and 1,200 respectively.

(c) The core supervisory cadre in Group 'C' was that of "Surveyor". Total strength of Surveyors may be kept around 800.

(d) Total strength of CCS Group 'B' should also be kept around 800.

(e) The structure of Group 'A' cadre may be kept in accordance with 1989 Rules. Total strength of CCS Group 'A' cadre may be kept 258. (This does not include the notional posts for MO GSGS or Geography Section General Staff).

(f) Cadre structure of various GCS Group 'A' 'B' and 'C' posts may be reviewed and redistributed by keeping in view the modernization aspects of the department. The concept of *paperless office* may be given due considerations.

Topographer

The qualification for Topographer must be Graduate (with Mathematics). The other features were as follows:

i) On recruitment as TTT'B' he should be placed in the pay scale of Rs 4,000- 100-6,000.

ii) On classification after 2 years, he may be placed in Topographer Grade Ill in pay scale of Rs 4,500-125-7,000.

iii) On completion of five years' service in Topographer Grade III and after passing departmental trade test, he may be placed in Topographer Grade II in the pay scale of Rs 5,000-150-8,000.

iv) On completion of five years' service in Topographer Grade II and after passing departmental trade test, he may be placed in Topographer Grade I in the pay scale of Rs 5,500-175-9,000.

Promotion avenues:

i) About 25 per cent posts of Group 'B' to be filled up from Topographers' cadre.

ii) After putting 5 years of regular service in the scale of Topographer Grade-I, he would be eligible for fast-track promotion by LCDE to group 'B' post.

Cartographer

The qualification for Cartographer was to be - B.Sc. or BCA or BA (Geography with mathematics at Intermediate Level)

i) On recruitment as TTT'B' he should be placed in the pay scale of Rs 4,000- 100-6,000.

ii) On classification after 2 years, he may be placed in Cartographer Grade III in the pay scale of Rs 4,500-125-7,000.

iii) On completion of five years' service in Cartographer Grade III and after passing departmental trade test, he may be placed in Cartographer Grade II in the pay scale of Rs.5,000-150-8,000.

iv) On completion of five years' service in Cartographer Grade II and after passing departmental trade test, he may be placed in Cartographer Grade I in die pay scale of Rs.5,500-175-9,000.

Promotion avenues:

i)   About 15 per cent posts of Group 'B' was to be filled from Cartographers' cadre.

ii) After putting 5 years of regular scale in the scale of Cartographer Grade-I, he would be eligible for fast-track promotion by LCDE to group 'B' post.

Surveyor

The qualifications for Surveyor must be - BE or B. Tech or BIT or MCA or M. Sc. (Mathematics, Physics).

i)   On recruitment as TTT 'A' he should be placed in pay scale of Rs 5,000-150-8,000.

ii) On classification after 2 years, he may be placed in scale Rs 5,500-175-9,000.

Promotion Avenues:

(a) On completion of five years, he will be eligible for promotion to Group 'B'. 35 percent posts of Group 'B' were to be earmarked for surveyors.

(b) On completion of 12 years of service he will be eligible for ACP in scale of Rs.6,500-200-10,500.

(c) On completion of 24 years of service he will be eligible for ACP in scale of Rs. 10,000-325-15,200.

Group 'B' Cadre

(a) The Group 'B' officers were to be designated as Officer Surveyor or Assistant Superintending Surveyor as per the scale. Recruitment to Group 'B' posts was to be through promotion from feeder Group 'C' cadres as indicated below:

- 25 per cent posts was to be filled up from Topographers' cadre.

- 15 per cent posts was to be filled up from Cartographers 'cadre.

- 35 per cent posts was to be filled up from Surveyor's cadre.

- 25 per cent posts was to be filled up through LDCE. All officers who have completed 5 years of regular scale in the scale of Topographer Grade-I or Cartographer Grade-I or Surveyor was to be eligible for fast-track promotion.

Nearly 50 per cent of the total Group 'B' posts were proposed to be in the higher scale of Rs.7,500-250-12,000 and designated as Assistant Superintending Surveyor.

Reorganization Proposal 2

One proposal received was concerning the duties and responsibilities of different officers at the Surveyor General's Office only in Dehradun. Details for each post was suggested and the same was almost followed as well. The whole gamut of activities in SGO can be gathered from the details given below.

I. Additional Surveyor General

1. Overall responsibilities for coordination and execution of Surveyor General's policy on all technical and administrative work of the Department.

2. Function as Additional Surveyor General for MPO, G&RB, DMC (D. DUN), DMC (Hyd.), STI, D. Survey (Air) and R&D.

3. Policy matters concerning collaboration with firms like Vandana Aviation etc. for formation of strategic alliances.

4. Screening and selection for deputation and training of personnel in India and abroad.

5. Deputizing for Surveyor General of India in various conferences and meetings etc. when ordered by him and in his absence on tour or leave.

6. All matters concerning restructuring of department in consultation with Surveyor General.

7. Matters related to aid programmes and international co-operation.

8. Exercise of statutory powers of Additional Surveyor General of the zones where the charge is held on current duty basis.

9. Other technical and administrative work assigned by Surveyor General from time to time.

10. National Spatial Data Infrastructure (NSDI) related activities.

11. Technical programme and review.

12. Digital policy for Survey of India.

13. Policy matters regarding digital centres and printing offices.

14. Specialized projects with various ministries.

15. Technical correspondence with ministries regarding restriction policy.

16. Processing of plans schemes and monitoring of approved schemes including sanctions and achievements.

17. Projection of requirement of extra-departmental surveys to zones and directorates.

18. Field and recess programmes of zones and directorates of Survey of India.

19. Control work of publications including annual reports and performance audit.

II. Deputy Surveyor General-I or DSG-I

1. Matters connected with aerial photography and attendance at various meetings and conferences.

2. Correspondence regarding pricing and copy right of digital products, topographical maps and Indian Air Force (IAF) maps.

3. Matters concerning participation of Survey of India in various exhibitions, seminars and conferences.

4. Matters connected with loss or destruction of publications.

5. Matters connected with survey priority conferences.

6. Liaison with Army Headquarters, Engineer-in-chief and MO GSGS on survey matters.

7. Other technical and administrative work assigned by Additional Surveyor General and Surveyor General from time to time.

8. All matters including disciplinary cases concerning Group 'A' Officers in consultation with Surveyor General.

9. Appellate authority for Group 'C' staff in SGO for punishment under CCS (CC&A) Rules, 14.

10. Review of retention in service under FR 56 (J) after completion of 30 years of service or 50 years of age.

11. Grant of medical advance as per medical advance rules, legal bills.

12. Correspondence regarding pay commission in respect of Group 'A' officers.

13. Disciplinary cases pertaining to Group 'B' Officers.

III. Deputy Surveyor General-II or DSG-II:

1. Parliament questions and administrative matters.

2. DPCs and ACPs connected with Group 'B' 'C' & 'D' employees of the Department and representations thereon.

3. Controlling officer of Surveyor General's Office (SGO) including Boundary Cell.

4. Disciplinary authority for Group 'C' staff for SGO under CCS (CC&A), Rules 14.

5. Controlling officer of DSOS and his staff.

6. Co-ordination of staff associations and departmental councils.

7. Cadre review of Group 'A' and 'B' officers.

8. Hindi liaison officer of the Department.

9. *Public Grievances Officer* of the Department.

10. Other technical and administrative work assigned by SG/Addl.SG from time to time.

11. Streamlining of O&M procedures and control of Work Study Unit.

12. Legal cases

IV. Production Technologist

1. Transfer/posting and other matters of Group 'B','C' and 'D'.

2. Procurement of stores, instruments and equipment for non-plan budget and issue of related sanctions etc. on behalf of S.G.

3. Movement and distribution of departmental equipment, machinery and transport.

4. Policy regarding condemnation or replacement of plan and equipment.

5. Advance for conveyance, HBA, advance from GPF and final withdrawal of other than Group A officers.

6. Other technical/administrative work assigned by SG/Addl. SG from time to time.

V. Deputy Surveyor General-III or DSG-III:

1. Security matters of the Department.

2. Fixation of pay and other matters of army officers.

3. Duties of Assistant Surveyor General

4. Pay and anomaly cases of Division-I and above.

5. Other technical/administrative work assigned by Surveyor General from time to time.

6. Digitization of cartographic databases.

7. NSDI node for SOI.

VI. Director (Administration & Finance) or DAF:

1. Vigilance cases.

2. Financial matters pertaining to the civil works.

3. Control of budget of the Department including appropriation and re-appropriation both for plan and non-plan scheme.

4. Audit objections and audit inspection reports.

5. Processing and monitoring cases regarding purchase of land, hiring of buildings, construction work and matters connected with survey estates at various location for Survey of India.

6. Foreign exchange cases.

7. To conduct enquiries in cases specially ordered by the Surveyor General.

8. Parliament questions and financial matters.

9. Pay fixation cases.

10. Recruitment rules of Division-I and above.

11. Zero base budget review of Survey of India.

12. Other technical and administrative work assigned by Surveyor General from time to time.

13. Advance for conveyance, HBA, advance from GPF and final with-drawal of Group A officers.

14. Great Trigometrical Survey or GTS related activities

VII.Deputy Surveyor General IV

1. Legal cases including appeal on disciplinary cases of Group 'C' staff of the Department except of Surveyor General's Office, Dehra Dun.

2. Schedule caste and schedule tribe (SC//ST) Liaison Officer of the Department

3. Any other duties as allotted by the Surveyor General of India/Additional Surveyor General from time to time.

Reorganization Proposal – 3

National Geospatial Data Centre

It was proposed to have a National Geospatial Data Centre (NGDC) in Dehra Dun with Surveyor General of India as the chief executive. NGDC was to directly control all state GDCs, and the proposed Indian Institute for Geospatial Information (IIGI) and National Institute for Geodesy & Earth Sciences (NIGES). IIGI and NIGES were suggested to be autonomous institutions with the Surveyor General of India as chairman and to have independent staff recruited as scientists. SOI was also to lend its officers on tenure basis. IIGI was recommended to be actively associate itself with universities and subsequently to be declared as *deemed university* authorized to give degrees including Ph.D. IIGI was also to conduct regular programmes for students and specific training programmes for the staff of SOI.

The NGDC was to have various groups *viz.*, data acquisition, products, business & finance, sales and marketing, solutions, and human resource development - each to be headed by an officer of the rank of surveyor general and supported by other staff. All officers of the rank of surveyor general will be members of the governing board was to be chaired by the Surveyor General of India. The steering committee for National Geospatial

Data Committee (NGDC) was proposed to have the Surveyor General of India as member secretary with deputy chairman, Planning Commission as the chairman and secretaries of concerned departments as members. The total staff was estimated to be 121: Surveyor General of India-1; Surveyor General-2; Additional Surveyor General-3; Directors-5; Superintending Surveyors (SS) or Deputy Directors (DD)-12; Group 'B'-16; Group 'C' Surveyors-22; and Group 'D'-60.

State Geospatial Data Centre (GDC)

Each state was proposed to have a geospatial data centre at the capital city of the state. GDCs were to be responsible all data in that state *viz.*, topographical, GIS, LIS, census, attribute data etc. The state chief was also to be called *Surveyor General*. All survey operations were be privatized as by then the industry was expected to become mature over a period of time. A Surveyor General in the states was to be assisted by two additional surveyor generals; one of them was to look after technical aspects and the other for contracts.

Each additional surveyor general was to have two officers of the level director; 6 officers of the level of SS or DD; and ten officers at the level of Surveyor or Deputy Superintendent Surveyor (DSS). In addition, each officer was recommended to have one personal assistant, one assistant and one messenger. Officers of the level additional surveyor general and above were expected to have one additional personal secretary. Thus, the strength of each major or minor state GDC was to be as below (Table 3.4).

Proposed posts	Major state	Minor state
Officers	39	20
Surveyors	100	60
Private secretary	3	2
Personnel assistant	78	40
Messenger	39	20
Total	259	142

Table 3.4 *Proposed strength of GDCs*

The total strength of SOI including the NGDC and the state GDCs was estimated to be 6,601. Further, persons from various disciplines having B.Tech. or M.Sc. qualifications were proposed to be recruited from the *Indian Geospatial Service* (IGS). Persons with B.Sc. qualification with mathematics as subject were to be taken as surveyors as group 'B' officers who were to rise up to the level of director. About 10 per cent of posts was to be filled up by lateral entry up to the level of additional surveyor general. The proposed pay-scales were as follows (Table 3.5):

Level	Scale in Rs.	Remarks
Entry level/ DSS	8, 000 - 13,000	-
Senior Time Scale I	10,000-15,200	After 4 years of service
Senior Time Scale II	12,000-16,500	After 9 years of service
Junior administrative grade/ director	16,400-20,000	-
Senior administrative grade/ additional surveyor general	18,400-22,400	-
Higher administrative grade/ surveyor general	22,400-26,000	-
Surveyor General of India	26,000	Fixed

Table 3.5 *Proposed pay scales*

All promotions up to the level of additional surveyor general was proposed to be based on *Flexibility Complementing Scheme* (FCS) and rest of them were to be vacancy-based recruitment. Attractive voluntary retirement schemes, sabbatical leave or deputation to government departments and even private sector was to be permitted. Recruitment of personnel was planned to be through *Indian Geospatial Services* (IGS). Further, most of the services was to be privatized such as transport, conservancy, map printing, data acquisition, quality control, and marketing. The total budget of SOI was about Rs 160 crores with a revenue of about Rs 15 crores. If the strength was to be brought down to estimated 40 per cent, thence here was to be a substantial savings of about Rs 40 crores. In addition, due to introduction of cost recovery mechanisms, a break-even stage was to be reached much faster than expected five years.

Reorganization Proposal 4

The induction of Group 'A' officers from Army was stopped in the past 18 years because of different reasons. Due to slow recruitment at lower levels especially in the technical offices, the production was affected adversely. The onus for this lies on the indifferent attitude of decision makers. The then scenario can be comprehended after analyzing the manpower status during the years from 2001 to 2005 (Table 3.6).

Cadre	Group 'A'	Group 'B'	Group 'C'	Group 'D'
Sanctioned Strength	315	627	7,418	7,917
01-01-2001	219	475	5,992	5,910
01-01-2002	196	421	5,616	5,634
01-01-2003	172	365	5,262	5,274
01-01-2004	159	310	4,802	4,861
01-01-2005	151	283	4,206	4,474

Table 3.6 *Depletion of manpower over next 5 years*

Note: There are about 90 different trades

In the proposed structure, the manpower was to be reduced from 16,271 to 9,9192 in five-year time. The strength in Group A was expected to increase from 315 to 811, *i.e.,* a change of +170 per cent. On the other hand, the Group D was to reduce from 7,917 to 2,960 (-70%). Further, in other cadres, changes were also proposed. However, the expenditure was kept almost same. Nevertheless, this attempt has been made similar to *Cadre Review* exercise in the Government of India.

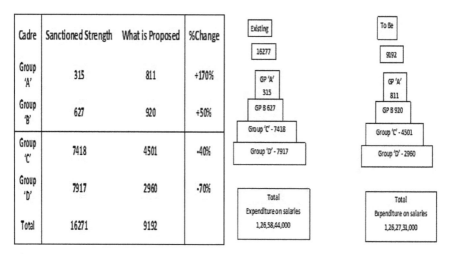

Cadre	Sanctioned Strength	What is Proposed	%Change
Group 'A'	315	811	+170%
Group 'B'	627	920	+50%
Group 'C'	7418	4501	-40%
Group 'D'	7917	2960	-70%
Total	16271	9192	

Existing

16277

GP 'A' 315
GP B 627
Group 'C' - 7418
Group 'D' - 7917

Total
Expenditure on salaries
1,26,58,44,000

To Be

9192

GP 'A' 811
GP B 920
Group 'C' - 4501
Group 'D' - 2960

Total
Expenditure on salaries
1,26,27,31,000

Figure 3.5 *Proposed restructuring*

In photogrammetric field, for data collection and updates from aerial photographs and imagery, the latest soft copy technique was then still to make an entry into SOI, whereas other departments and institutions such as National Remote Sensing Centre (NRSC), Indian Institute of Remote Sensing (IIRS), Advanced Data Processing Research Institute (ADRIN) and the like have it since past several years. Almost all other central government technical organisations have switched over from analogue to digital system in their technical functioning, but SOI had introduced it in some offices partially only. It was then predicted that SOI will take few more years to change over to digital environment even if the procurement of computer system and software continues. However, the changeover to digital environment was considered to be a futile exercise if it did not commence with the recruitment of technical personnel in a phased manner at the earliest. The existing employees were at higher age groups and the rate of their retirement was so large that many offices may have to be closed down in near future. In fact, the process of winding up of some offices (survey parties) had already begun.

Then the organisation had more than ninety trades some of which were obsolete, some isolated and some were redundant because of change in technology and methodology. Manpower was becoming expensive; it became extremely important for any organisation to circumvent inactivity of manpower. Therefore, it was suggested that a single trade of *surveyor* may replace about 90 different trades. The surveyor was proposed to be the workforce who was to carry out all types of technical and administrative work. These surveyors should be given thorough training in surveying and mapping for two years initially at their entry to SOI. Before posting them to any specialized job such as map reproduction, store, record keeping etc., they should be given short training ranging from one to three months. With the then structure, technology and manpower profile, SOI was not geared to meet the demand for value-added products. Apart from infusing new technology, SOI was required re-engineer its work procedures, retrain the existing personnel and augment technical manpower. Hence, the need for restructuring and manpower augmentation was proposed keeping in view of the then departmental activities:

I. Provision of control framework including gravimetric and geomagnetic, and tide prediction.

II. Development and maintenance of National GPS Data Centre for dissemination of data through NSDI.

III. Processing and analysis of geodetic and geophysical data for tectonics studies.

IV. Topographic mapping in restricted areas and other classified mapping.

V. Cartographic product development and interaction with users.

VI. Creation of national geospatial data Library.

VII. Creation and maintenance of value-added geospatial products, creation and maintenance of LIS or GIS data bank

VIII. Printing of classified maps.

IX. Certification of external boundaries appearing on maps published by SOI as well as other agencies.

X. Training in surveying and mapping and R&D activities.

XI. Quality control and certification of geospatial data collected through out-sourcing.

The outsourcing of technical work included components of all unrestricted surveying and mapping work; project surveys; special maps, geographical maps and printing of *Open Series* maps and other unrestricted maps and tidal data collection. Restructuring was considered to be linked with human resources development by taking following measures:

- Screening of the existing manpower for re-deployment into new role after training.

- Periodic upgradation of skills through refresher courses.

- Defining work norms and providing transparent productivity linked incentives as a major tool to increase efficiency and productivity.

- Ensuring adequate promotional avenues as an incentive for career enhancement.

- Rationalizing the existing trade structure to create a cohesive cadre. Inducting new personnel to make up the deficiency for fulfilling the modified mission.

- Modifying the existing recruitment rules of Group A service to create a unified cadre (*Indian Survey Service*) which was to support the needs of SOI as well as Military Survey.

- Implementing *Flexible Complementing Scheme* up to Scientist F level.

- Reducing the number of technical, store and record keeping trades in Group C service from existing 90 to just one.

The other feature of restructuring was related to the functioning of the organisation which included following steps:

- Creation of more directorates - more or less state wise.

- Consolidation of spread-out parties into directorates at one location. Location of directorates in their area of responsibilities, mainly state capitals.

- Creation of *Indian Survey Corporation Limited* (ISCL).

- Creation of National Spatial Data Infrastructure (NSDI).

• Each region to have modern printing facility.

• Introduction of state-of-the-art technology in all fields.

Educational Qualifications

Apart from man-power and structural reorganization, educational qualification was considered important with changing technology and market
requirements. Following educational background was proposed:

Group A: Engineers or post graduate either in mathematics or
physics or geophysics or computer science or surveying or geodesy
or photogrammetry or remote sensing or geomatics or cartography
with compulsory mathematics at graduation level.

Group C: B.Sc. or BA having mathematics or computer sciences
with mathematics as one of the compulsory subjects.

The recruitment for Group A was proposed to be initiated immediately
with the new qualifications. However, for Group B, promotion from Group
C should be started. Further, regarding Group C, the suggestions were as
follows:

(a) Some percentage of the staff was proposed to be recruited per year
during transition period,

(b) Redeployment of existing Division II staff through suitable selection
procedure

(c) Phasing out of surplus personnel,

(d) Surplus cell or Voluntary Retirement Scheme or VRS should be considered. Hence, no new recruitment.

Regarding Group D, no new recruitment was proposed. The proposed
structure was as shown in Table 3.7.

Group 'A' Existing	Proposed	No.	Level-Scale of Pay
Surveyor General of India	Surveyor General of India	1	S32 – Rs. 24,050-650-26,000
Additional Surveyor General	Additional Surveyor General	14	S30 – Rs. 22,400-525-24,500
Director/Deputy Surveyor General/ Deputy Director (Selection Grade)	Director	25	S29 – Rs. 18,400-500-22,400
	Joint Director	-	S26 – Rs. 16,400-450-20900
Deputy Director	Deputy Director	-	S24 – Rs. 14,300-400-18,300
Superintendent Surveyor (Selection Grade)	Superintendent Surveyor (Selection Grade)	-	S21 – Rs. 12,000-325-15,200
Superintendent Surveyor	Superintendent Surveyor	-	S19 – Rs. 10,000-325-15,200
Deputy Superintendent Surveyor	Deputy Superintendent Surveyor	100	S15 – Rs. 8,000-275-13,500

Table 3.7 *Existing and proposed structures*

Flexibility Complementing Scheme or FCS from DSS (Scientist B) up total of Joint Surveyor General + DD+ SS (Selection Grade) + SS was proposed but not to exceed 427. Promotions at all levels were also proposed including three levels in Group C, *i.e.*, Surveyor, Senior, Surveyor and Chief Surveyor to Group B or Officer Surveyor. Further, the major ingredients in restructuring or re-engineering were also identified. Then the major strength was deliberated to be honesty and integrity of its employees which was proposed to be exploited and further strengthened. In this regard, the following steps are suggested:

(a) Faithful implementation of government rules and regulation and faster redressal of grievances of individual or group of employees. Number of court cases in SOI were on the rise mainly due to inaction on the part of decision makers at various levels. It appeared that the officers and staff were compelled to go to the court for redressal.

(b) Government must understand the basic difference between the functioning of SOI and other departments with respect to commercialization and downsizing of department. The then policy of decision makers not allowing recruitment of technical personnel was considered to be risky which was causing irreparable damage not only to SOI but also to the nation. It was realized that the work of SOI was such that even a highly qualified person takes at least 5 years before becoming productive.

(c) Furthermore, only certain portion of SOI's work can be commercialized or given to outside agencies. Therefore, it was felt necessary to bring out correct perspectives as far as downsizing and commercialization of SOI were concerned.

Other measured included placement of right person at right place, timely promotions, proper motivation, and upgradation of grades and amendment of recruitment rules.

Reorganization Proposal 5

The officers of Survey of India are on deputation from the Ministry of Defence or specifically from MO GSGS, or from the Indian Engineering Service or by promotion. There were always different views about the cadres, their promotion and service conditions. However, the Surveyor General of India has been a civilian post which most of the time occupied by army officers. The status of Defence stream has been shown in Table 3.8.

POST	Sanctioned			Posted			Vacant	Remarks
	SOI	Mil. Svy	Total	SOI	Mil. Svy	Total		
Surveyor General of India	-	-	-	-	-	-	-	
Additional Surveyor General	3	1	4	6	1	7	(-) 3*	
Director	32	25	57	20 (+)3*	18	41	16	
Deputy Director	28	23	51	-	1	1	50	
Superintending Surveyor	28	44	72	2	-	2	70	25 posts are to be operated at lower level and 13 posts temporarily transferred to Civil Stream
Deputy Superintending Surveyor	-	4	4	-	-	-	4	70 posts are to be revised

Table 3.8 *Group A: Defence Stream as on 29-10-2003*

There were several posts lying vacant for different reasons. The largest vacancies were at the level of Superintendent Surveyor (SS) and Deputy Director (DD). How and from which stream these posts were to be filled up had always been a bone of contention. The court orders have further complicated the situation. As a result, posts were not filled up. Hence, there has been proposal to amendment of recruitment rules specifically for Group A posts. The proposal for reorganisation included following suggestions:

(a) To include more disciplines so as to make broad based recruitment of officers in Group 'A' Service.

(b) To delink SOI from Engineering Service.

Department of Personnel and Training or DOP&T and Ministry of Railways have given their 'no objections' to the delinking of SOI from Engineering Services Examinations. Union Public Service Commission (UPSC) had also agreed to fill up these vacancies on direct recruitment basis. Further, amendment to the recruitment rules had been proposed by SOI and submitted to DST to include more disciplines such as degree in engineering or M.Sc. in Geography or Geology or Computer Science or Physics or Mathematics or Geophysics.

Amendment to Recruitment Rules for Group 'A' post in Survey of India

Induction of army officers has been approved by the Minister for Science & Technology and Hon'ble Raksha Mantri. However, a large number of posts are lying vacant at SS and DD level as well. Engineer-in-Chief's branch offered to send officers at the lower level *i.e.*, Major, Lt Col (Time Scale) or Lt. Col. Further, there has been a provision in the existing recruitment rules to fill up the posts by transfer on deputation of officers holding analogous posts in *Corps of Engineers* who have earlier worked in SOI and undergone initial training of two years. The proposal was to have period of deputation in following way: Senior Administrative Grade: 5 years; Non-Function Selection Grade: 5 years; Junior Administrative Grade: 4 years; and Scientific and Technological Services (STS): 3 years.

Since a number of posts were lying vacant in SOI which was creating functional difficulties, it was recommended that the matter may be taken up with the Ministry of Defence (MOD) to induct suitable officers on deputation after carrying out proper interviews. However, a need was felt to get earlier recruitment rules modified to the extent that two years training may be relaxed to short survey course, which is being conducted by College of Military Engineering, Pune. These officers were to be given   training of two months in Survey Training Institute regarding various procedures and operations being carried out in terms of administrative and technical responsibilities. In order to circumvent this situation, it was proposed that 70 posts at the level of Deputy Superintendent Surveyor (DSS) of Defence stream was to be revised and induction may be initiated from the army. Further, the situation of Civilian Stream of Group A post has been shown in Table 3.9.

Post	Sanc-tioned	Posted	Vacant	Remarks
Surveyor General of India	1	1	-	
Additional Surveyor General	3	-	3	DPC held
Director	33	17	16	DPC held
Deputy Director	37	17	20	
Superintending Surveyor	50	44	6	
Deputy Superintending Surveyor	42	11	31	

Table 3.9 *Group A: Civil stream as on 29-10-2003*

All these five proposals were prepared by different groups and cadres. Hence, they have highlighted to issues relevant to them. It was difficult to resolve all the issues in one attempt. Hence a step-by-step action seemed to be more acceptable. Nevertheless, the process has to be mooted. Reduction of staff, particularly Group D, was generally agreeable. Recruitment rules, their implementation, posting and promotions are other issues related to reorganization. In some proposals, a separate 'service' for SOI was proposed. Further, there were different views or even no comments regarding links with the *Indian Engineering Services* or deputation and reverse deputation from the army. Nevertheless, an attempt was made in this direction keeping in view of not only the cadres, but also smooth functioning of the organisation and better acceptance of the technology.

Restructured Survey of India

Due to such initiatives, there was a serious thinking to transform the organisation, including establishment of a commercial arm of SOI. The other achievements were also path-breaking and in some cases a major step forward in the administrative and technical performance. Some of the achievements were:

- All 1:50,000 scale topographical maps were digitized in house.

- Initiatives for seamless data on new datum taken.

- Control points for WGS 84 series were almost completed.

- Data already provided to National Resource Information System (NRIS) programme of the Department of Space (DOS)

- Updating of topographical sheets with remote sensing data and aerial photographs using modern fieldwork techniques.

- All the geospatial data centres (GDCs) were equipped with PCs, WS, GPS, software in order to make them self-sustaining.

- Printing was decentralized.

- Role of G&RB Directorate has been changed.

- STI is underwent changes: (a) the courses are made shorter, and (b) modern technology was introduced.

- ATLM experiment was conducted, results were evaluated.

- Preparation of large-scale maps at 1:2,000 scale was initiated for towns.

Considering the compelling demand for change in view of depleting manpower, changing technology, new applications based on geospa-tial data, it became necessary to reorganize the whole set up. The major transformation was required in the office of the Surveyor General (SGO) itself. The basic task shifted from the production of maps to deliver digital geospatial data. Another priority was to include development-related issues in the functioning of the organisation by working closely with the state governments spread all over the country. Further, new directorates were opened up keeping in view of the new vision. Some of them were Research & Development, Business & Publicity, National Geospatial Data Centre and National Spatial Data Infrastructure. The reorganized structure was to be more functional and affective (Fig 3.6).

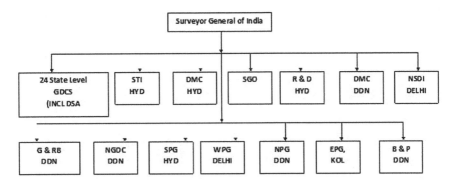

Figure 3.6 *Restructured setup*

Due to the above steps taken, there were certain advantages which had a long-lasting impact in the functioning of the organisation. It became possible to directly monitor the technical work at GDCs. All individuals are made responsible with assigned targets. Hence, there was a quick response at all levels, particularly from GDCs to SGO and *vice-versa*. The new setup with the geospatial data centres was as follows (Fig 3.7).

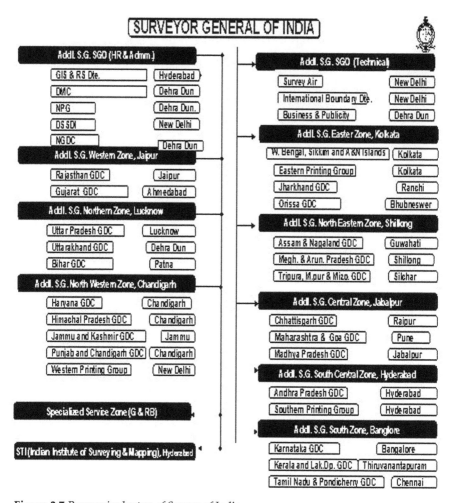

Figure 3.7 *Reorganized setup of Survey of India*

Opening of Tamil Nadu GDC and Other Directorates

The Great Trigonometrical Survey was initiated from St Thomas Mount near Chennai airport. This was a major scientific initiative of that time, but there was no Survey of India office in the whole state of Tamil Nadu. Hence, the then Secretary of the Department of Science & Technology and the Surveyor General of India and his team met the Chief Minister of the state and were assured for a piece of land (Fig 3.8). However, initially the Tamil Nadu Geospatial Data Centre was established in a pool central government premises in Chennai.

Special attention was given to northeastern India as well. The erstwhile directorate in Shillong (Meghalaya) was reorganized into three GDCs. The states and union territories which were smaller in size were coupled with the nearby states, such as Gujarat with Daman & Diu; or West Bengal with Sikkim and Andaman & Nicobar Is.; or Kerala with Lakshadweep. Further, printing establishments were seperated from the activities of the GDCs. Hence, new printing directorates at Kolkata, Hyderabad, Dehra Dun and Delhi were raized. This change gave better opportunities for promotions of senior officers. Furthermore, the establishment of National Spatial Data Infrastructure directorate was also formalized. The Boundary Cell continued to be attached with the Surveyor General's Office (SGO).

CM promises all help to set up Survey of India office

By Our Special Correspondent

CHENNAI, SEPT. 17. The State Government will provide all official assistance for establishing an office of the Survey of India, the Chief Minister, Jayalaithaa, today told a delegation from the Union Department of Science and Technology.

The team led by the department Secretary, V.S. Ramamurthy; the Surveyor-General of India, Prithvish Nag; the Director, National Spatial Data Infrastructure, R. Shiva Kumar, and the Anna University Vice-Chancellor, E. Balagurusamy, met the Chief Minister mainly to brief her about a Centre for Geo Spatial sciences at the Anna University.

A Chair (named after India's

first Surveyor-General Col. Mckenzie) had been established in the University's Institute of Remote Sensing with Rs. 50 lakhs provided by the DST.

It would train teachers in the specialised discipline, first such venture in the country, an official release said.

The Survey of India did not have an office in Tamil Nadu, the delegation said and sought the Government's help in finding the required land.

The Chief Minister, according to the release, assured the delegation of all help in this regard.

Prof. Ramamurthy said the department proposed to work with the Anna University on a project on women's empowerment, for training technically qualified women. With the

training, women could take up profitable employment. The release said the Chief Minister expressed happiness at the venture and offered to support its implementation in the State.

Earlier in the day, Prof. Ramamurthy declared open the Spatial data centre at the university and handed over a cheque for Rs. 50 lakhs for establishing the Mckenzie Chair at the University's Institute for Remote Sensing.

He said spatial data creation and use of such data in modern areas held great potential for both private enterprise and the public sector.

Prof. Balagurusamy said provision of infrastructure should necessarily begin with planning and spatial data were a prerequisite for such planning.

Remembering a master cartographer

By K. Ramachandran

P. Nag (extreme right), Surveyor-General of India, after unveiling a bust of William Lambton on Wednesday. Others seen from the left are V.S. Ramamurthy, secretary, Department of Science and Technology, Om Kumar, Chief Executive, Tamil Nadu Technology Mission, and E. Balagurusamy, Vice-Chancellor, Anna University. — Photo: K. Pichumani.

CHENNAI, SEPT. 17. A brief function held at St. Thomas Mount in Chennai on Wednesday, to unveil a bronze bust of William Lambton, symbolised a visit to another era. An era 200 years back, when Lambton used a nicely constructed Theodolite and the raw courage of a band of men for the greatest survey and mapping work ever attempted. And finally succeeded.

On April 10, 1802, Lambton began the Great Trigonometrical Survey (GTS) to completely measure and map the Indian subcontinent by measuring a 12-km long baseline near Madras (now Chennai). The baseline was needed to form The Great Arc — a survey comprising a series of measurements running from India's North to South — with utmost accuracy.

For the next 50 years, Lambton and his successor George Everest consolidated and coordinated the stupendous task. They used the baseline to construct and measure a series of triangles. Having connected the two sides of the peninsula, Lambton devoted much of his labour to the measurement of an arc of the meridian. Also the web of triangles was

extended to establish the position of main cities and centres.

The survey team traversed the most

hostile terrain and populations, and lost more lives than in any war. The triangulation system built up by Everest also

made it possible to discover the world's highest mountain, which appropriately was named after him.

Prithvish Nag, now the Surveyor-General of India, who unveiled Lambton's bust, places the work of the GTS in today's perspective thus: The Great Arc is the foundation on which the Survey of India was now building on. Almost all of India's development work has benefited from the GTS. Today's function recalls the missionary zeal of the Survey of India's first participants to map the country.

An entourage of the Union Department of Science and Technology is in Chennai to wrap up the celebrations organised to mark the 200th anniversary of the Great Arc. The Department Secretary, V.S. Ramamurthy, who launched a Tamil Nadu Geospatial Data Centre at the Anna University's Institute of Remote Sensing, said Chennai had a special place for the occasion. It was in this city the 12-km baseline up to Marina beachfront was formed to begin the survey. Secondly, Chennai was home to India's first School of Survey. This school later turned into the College of Engineering, Guindy, and now has taken the shape of Anna University.

Figure 3.8 *New Tamil Nadu GDC*

The Tamil Nadu and Pondicherry Geospatial Data Centre started functioning in Chennai on 24 February 2004 from Anna University campus. This centre shifted to a rented premises in the state owned SIDCO Electronic Complex at Guindy on 17th June 2004. It was entrusted to prepare 44 topographical sheets at 1:50,000 scale with field verification during the then field season beginning from October 2004.

Reorganization of North-East India

Special attention was given to the whole of northern eastern India then having seven states. Earlier, the survey matters for this region were dealt by its North Eastern Circle located at Bonnie Brae Estate in the heart of Shillong city and partly by the erstwhile survey headquarters in Kolkata (Woods Street) due to historical reasons. In Shillong, there were two more offices in Dhankheti and Bomfyle Road. The other offices were located in Silchar (Ambicapatti) and Guwahati (Ganeshguri Chariali); while No. 22 (P) Party was situated in Dehra Dun (17 E.C. Road). Further, a hillock at the out-skirts of Guwahati city was in possession of the SOI as well.

Assam Governor Lt Gen (retd) Ajai Singh showing a map during the inauguration function of Assam and Nagaland Geo-Spatial Data Centre at Srimanta Kalakshetra, in the city, on Tuesday. (Senti)

Figure 3.9 *Inauguration of Assam-Nagaland GDC in Guwahati*

Considering the survey and mapping requirements of this region and also the diversities, it became essential to reorganize the whole setup in the following manner:

(a) Assam & Nagaland Geospatial Data Centre at Guwahati having a wing in Shillong.

(b) Meghalaya & Arunachal Pradesh Geospatial Data Centre at Shillong.

(c) Tripura, Manipur & Mizoram Geospatial Data Centre at Silchar having a wing in Shillong.

The formal inauguration of the Assam & Nagaland Geospatial Data Centre at Guwahati was held on 31st March 2004 by the then Governor of Assam, Lt Gen Ajai Singh (Fig 3.9). The objective was to have infrastructure for acquisition, transformation and management of geospatial data. Further, tit was expected to have an all-round development in the northeastern region including Assam and Nagaland. Apart from preparing the District Planning Map Series of some of the districts and guide maps, SOI was involved in several development projects, to name a few (a) Barak Watershed Management Survey, (b) Narangi Cantonment Survey, (c) Assam-Meghalaya Boundary Demarcation Survey, (d) Jorhat Airfield Survey, (e) Dinjan Airfield and Cantonment Survey, (f) Nagaland-Manipur Boundary Survey, (g) Chabua Airfield Survey, (h) Lumding-Badarpur Railway Alignment Survey, (h) Brahmaputra Basin Survey, (i) Bongaigaon Thermal Power Project, (j) Cadastral Survey of Nagaland and the like. Furthermore, due to the department's initiative to digitize all the 165 topographical sheets at 1:50,000 scale for both the states, such reorganization became necessary. The same is true for other states of this region. The whole objective was to work closely with state governments for their social and economic development. Further, in order to give more importance to the norther eastern part of the country, all India Additional Surveyor General's Conference was held in Shillong, Meghalaya from 21-22 July 2008. This event was organized with an objective of strengthening the transformation process initiated a few years earlier (Fig 3.10).

Figure 3.10 *Additional Surveyor General's Conference, 2008 in Shillong*

Overall Strategy

Considering the requirement of an efficient system, application of modern technology and working closer with the state government, a reorganized structure was implemented. The objective was to give priority to *development* in addition to *security*. Hence, it was necessary to understand the development and planning issues of different states or their parts. Later this set up became very handy when *Svamitva* project for computerization of land records was taken up. Though, in 1905, for the sake of *Modernization of Survey of India*, revenue survey was separated from topographical surveys.

Keeping in view of the changed priorities, SOI was to re-engineer its processes and bring effective organizational restructuring to suit the changing business strategy. It was considering building a culture of continuous innovation in all aspects of its working and to encourage an environment conducive to an entrepreneurial culture among its employees. Further, its units were to be made self-sufficient having full component of administrative, stores, records and technical capabilities. But they became unwieldy because of the reduced strength. In order to achieve greater flexibility and re-deploy the existing manpower to cater for the needs of the states, SOI was re-organized into state level geospatial spatial data centres (GDCs) with all components of modern methods of geospatial data processing and marketing.

The zonal offices became less relevant and only serve as post offices. It had only increased the channel of communication, thereby, the directorates, which were the ground level functionaries, lost freedom in discharging the duties and lost communication with the SGO as well. Delegation of power from SGO to zones did not take place. Hence, the zonal set up was abolished with greater delegation of power from the Surveyor General to directors in all spheres of administrative and financial functioning. Nevertheless, the SGO itself was also reorganized. The day-to-day affairs of procurement of equipment, training of officers and staff, promotions up to Group B level, transfers and the like were proposed to be dealt by the directors.

The units at various places in a state were regrouped into GDCs with available manpower and with all components of geospatial data processing. Depending on the size of the state and amount of workload, additional manpower for technical work and supervision was estimated by the directorate at local level to achieve greater flexibility, accountability and user responsiveness. Further, the GDCs were to be strengthened to become a complete solution provider. Towards this end, greater autonomy in financial, decision making, and administration authority was to be delegated. This was to introduce accountability of the scientific and technical work at GDC level.

The GDCs were to interact with the state governments to assess their needs. They were also to enter into MOU with the state governments so that the data be more useful for application and thereby helpful in the creation of national data base. There was an additional objective of creating national connectivity for topographical and even cadastral data base. On the other hand, SGO was to immediately establish a dedicated internet linking all the GDCs. It was to introduce video conferencing for quick and effective decision making. Each directorate was given the responsibility of establishing state level spatial data centres (SDI).

The segments of work which could be out-sourced earlier even with adequate supervision and quality control by SOI personnel, was to be made more effective now. Available manpower in different skills were re-trained and redeployed in the desired trades. Special set up for marketing of products, market assessment, human resource development and product generation were to be introduced in partnership mode or on contract basis. Even lateral induction of critical expertise was considered. In addition, at SGO a dedicated management team was to be instituted (3 to 4 officers) having full support of DST for re-engineering. This team, in consultation with GDCs, was to report directly to the Surveyor General of India. On the other hand, the SGO was to formulate guidelines, standards and monitor the progress of work. The restructured setup can be gathered form (Table 3.10).

Sl. NO.	Existing topographical circles	Restructured state-wise circles	Sanctioned strength of Addl. SG	Sanctioned strength of Director/ DSG
1	Northern Circle, Dehradun	1 Uttar Pradesh GDC, Lucknow	1	3
		2 Uttaranachal GDC, Dehradun	--	1
2	North Western Circle, Chandigarh	1 Punjab& Chandigarh GDC, Chandigarh	--	1
		2 Haryana GDC, Chandigarh	--	1
		3 Jammu & Kashmir GDC, Chandigarh	--	1
		4 Himanchal Pradesh GDC, Chandigarh	--	1
3	North Eastern Circle, Shillong	1 Assam & Nagaland GDC, Guwahati	--	1
			--	--
		2 Meghalaya & Arunachal Pradesh GDC, Shillong	--	1
		3 Tripura, Manipur & Mizoram GDC, GDC, Silchar	--	1
4	Eastern Circle, Kolkata	1 Bihar Geospatial GDC, Patna	--	1
		2 West Bengal, Sikkim, Arunachal and	--	1
		3 Eastern Printing, A&N Is.	--	1
5	South Eastern Circle, Bhubaneswar	1 Orissa GDC, Bhubaneshwar	1	2
		2 Jharkhand GDC, Ranchi	--	1
6	Central Circle, Jabalpur	1 MP GDC, Jabalpur	1	3
		2 Chhattisgarh GDC	--	1
7	Western Circle, Jaipur	1 Rajasthan GDC, Jaipur	1	3
		2 Gujarat GDC and Daman & DIU GDCs, Ahmedabad	--	1
8	South South Eastern Circle, Hyderabad	1 Andhra Pradesh GDC, Hyderabad	1	3
		2 Southern Printing Group, Hyderabad	--	1
9	South Central Circle, Hyderabad	1 Maharashtra GDC, Pune	1	3
		2 Gao GDC, Pune	--	1
10	Southern Circle, Bangalore	1 Karnataka GDC, Bangalore	--	1
		2 Tamil Nādu & Pondicherry GDC, Chennai	--	1
		3 Kerala & Lakshadweep GDC, Thiruvananthapuram	--	1
11	Survey (Air), New Delhi	1 Survey (AIR) & Delhi GDC, New Delhi	--	2
		2 Western Printing Group, New Delhi	--	1

Table 3.10 *Restructuring of topographical circles*

Restructuring of Specialized Directorates

Sl. No.	Existing directorates	Restructured directorates	Sanctioned strength of Addl. SG	Sanctioned strength of Director/ DSG
1	Modern Computer Centre, Dehra Dun & Digital Mapping Centre, Dehradun	Merged as Digital Mapping Centre, Dehradun	--	2
2	Digital Mapping Centre, Hyderabad	Digital Mapping Centre, Hyderabad	--	2
3	Survey Training Institute, Hyderabad	Survey Training Institute, Hyderabad	--	7
4	Research & Development, Hyderabad	Research & Development, Hyderabad	--	1
5	Geodetic & Research Branch, Dehradun	Geodetic & Research Branch, Dehradun	--	2
6	Business & Publicity, Dehradun	Business & Publicity, Dehradun	--	1
7	Map Publication Office, Dehradun	Northern Printing Group, Dehradun	--	1
8	National Spatial Data Infrastructure, Dehradun	National Spatial Data Infrastructure, Dehradun	--	2
9	Survey General's Office, Dehradun & SOS Dehradun	Merged as Survey General's Office, Dehradun	1	8

Based on Departmental Order No. W-1137/709-GDC, dated 19 December 2003.

The existing zonal offices also stand abolished. The staff and resources of these zonal offices were utilized for the restructured state geospatial data centres and were attached with the existing circles in the stations for pay and allowances and other administrative work. Further, all the field and photogrammetric units and drawing offices which were under the control of topographical circles and specialised directorates were also abolished by merging the sanctioned strength of these units into the newly restructured state geospatial data centres.

Chapter 4

Geodesy

The advancement of science and technology in India owes much to the right steps taken in the past. We have been reaping the harvest now. One of such efforts has been the initiation in the field of geodesy about two centuries earlier. This action has become the backbone of the scientific surveys in the country. India has been a pioneer in this scientific pursuit which has been a matter of pride as well. As mentioned in Chapter 1, science and technology have been a part of the great Indian civilization. Scientific temper of the country is manifested from the ancient literature and knowledge. The Great Trigonometrical Survey should be looked in the same sequence. The pains-taking efforts of the surveyors of the SOI have provided a footing for the science and technology related activities of the country when there was no possibility of immediate gains. Nevertheless, a knowledge-based empire has been built up. India has been in the forefront of this scientific activity as well. Further, the objective of the Department of Science & Technology has been to patronize the foundations of the scientific and technological capa-bilities of our country. Its primary role has been as a catalyst, as a facilitator and as an enabler.

Geodesy has been one of the activities of the SOI which has contributed to the scientific foundation of the country. The accurate locations, availability of resources, identification of the priority areas and the like have roots in this scientific activity. Hence, geodesy has an increasing and valuable role in planning and development, particularly when government is taking new initiatives such as smart cities or digital India. Three-dimensional planning requires an accurate reference point for making measurement in X, Y and Z directions. Since changes are continuously happening, this branch of science provides stable coordinate reference frame for the whole earth including India. Hence, Global Geodetic Reference Frame had become mandatory.

As the scientific temper of the country keeps on improving, the signifi-cance of accurate measurements has been felt. A step taken two hundred years earlier in Madras, now Chennai, to initiate accurate measurement of the country had created wonders in the mapping of India and the neighbouring countries. Even the most sophisticated technologies cannot ignore the scientific wealth collected over the years. Earlier, the revenue surveys and urban planning have been bases on such initiatives. The knowledge acquired had been extremely useful for application of remote sensing and the geographical information system. The modern telecommunication network is also based on similar information. The proficiency of 250 years of survey provides an occasion for introspection and to exemplify the future course of action by taking advantage from the past experiences. Its practical applications include navigation, building of roads, tunnels, laying of railways, construction of airports and other infrastructure, demarcation of international boundaries and real estate properties. With the application of space-based technologies, geodesy has become more significant. Precise location of points required integration of space technology with geodesy. Most earth system science studies have been benefited from detailed knowledge of changes in the earth's shape, gravity field, and rotations. The combined techniques have contributed towards (a) unprecedented accuracy, (b) reduced latency, and (c) extraordinary spatial and temporal resolution. Further, ground-based geodetic network has helped in the operation of Global Navigation Satellite Systems or GNSS. The potential of the space-geodetic techniques has been developed during the last two decades under the auspice of the International Association of Geodesy (IAG).

Geodesy means the scientific investigation of the size, shape and structure of the earth. The main purpose of measuring the arc of the meridian was to define the spheroid. There are a number of activities which were required to be suitably combined in order to get the exact shape of the geoid or spheroid based on the measurement of the great trigonometric arc. In 1830, Sir George Everest devised a spheroid to fit in with geoid over South Asia, popularly known as Indian subcontinent - which is acknowledged as Everest Spheroid with its axis passing through Kalianpur as centre. This is not only meant for India but is being used by Pakistan, Nepal, Myanmar, Sri Lanka, Bangladesh, Bhutan and other southeast Asian countries. Initially, this spheroid had some inherent weaknesses and some corrections had to be applied in order to make it more acceptable.

At present, the entire load of geodetic and geophysical work for the whole country is being entrusted to the Geodetic & Research Branch (G&RB) of

SOI located in Dehra Dun. A large number of demands for geodetic work like seismotectonics studies, scientific monitoring of tectonic movements, provision of geodetic plan and height control, redefinition of Indian geodetic datum, densification of gravity and geomagnetic work as well as requirements of developmental projects with regard to geodetic and geophysical surveys are being entrusted to this branch. Considering the importance of this field of science, it was even proposed to convert one topographical field party from each of the zones to geodetic field party to enable them to expeditiously undertake geodetic tasks falling in their areas of responsibility. One field party of Southern Zone already stands converted into a geodetic party. The distribution of geodetic and geophysical work to different field units was to ensure that G&RB was is free to concentrate on the much-needed scientific research projects lying pending because of lack of resources. Further, it was proposed to establish a centre for seismotectonics studies in SOI. This centre was to have lateral entry of scientists from various disciplines of earth sciences so that integrated spectrum of studies can be undertaken .

Earlier Initiatives

In 1802, Major William Lambton initiated the measurement of the Great Indian Arc of the Meridian accompanied with the Great Trigonometrical Surveys (GTS)(Fig 4.1). The measurements were spread over 1,600 miles in length along the meridian and 1,200 miles in latitude at places to cover the entire country. It took approximately fifty years to complete the task and has been hailed as most stupendous work in the history of science. It was the time when there were no proper roads and the mode of transport was animal driven carriages, and most of the people covered distances on foot carrying their loads on head. Indians were employed to do the spadework in this great task and bore the brunt of the initial attacks from insects, animals and others, including vagaries of weather and terrain. Like true soldiers, they suffered most in the process. It is no wonder that this complete operation cost more lives than most contemporary wars, and most of the surveyors who laid their lives for the cause were Indians.

Figure 4.1 *Bronze placed at St Thomas Mount, Chennai*

The Great Trigonometrical Survey

The Great Trigonometrical Survey or GTS in due course turned out to be a mega scientific activity not only for undivided India, but for the whole world *per se*. It laid the scientific basis for measuring the base line and thereby providing reliable data to find out the shape and size of the earth. The exercise was based on triangulation over the length and breadth of the country and hence it is known as GTS. The basic requirement for a map or even GIS has been to know the location or geodetic coordinates in terms of longitude, latitude and elevation above the sea level. A framework of control points was established throughout the country on which geographical or topographical features have been adjusted. This framework is being known as *triangulation network*. The chains of triangles whose lengths and sides are measured for converting the distances along the spherical surfaces of the earth to geodetic coordinates. A series of triangles were measured along the 78° E of Greenwich, *i.e.,* from Kanyakumari to the foothills of Himalayas at Banog. This chain of triangles is now known as *The Great Meridional Arc of India*. With the success of this activity, the surveying technique changed from astronomical to triangulation method which was more accurate and economical as well.

Ultimately, the chains of triangulation covered the whole country and provided a grid iron mesh of points which became the basis for topographical, revenue and cadastral surveys (Appendix II). This work which was initiated by Col. William Lambton and completed by Sir George Everest. This helped in calculating the parameters of the Everest datum or the Indian datum which were as follows:

Semi major axis (a) = 6377301.243m or 20922932 ft

Semi minor axis (b) = 6356100.231m or 20853375 ft

Flattening (f) = 1/300.8017

India has traveled quite far in last two hundred years. *i.e.* from the days of astronomical methods to GTS and finally to the Global Positioning System or GPS.

The Indian Datum

Since the earth surface is uneven and not exactly suitable for direct mathematical computations, a hypothetical reference surface was to be defined which was known as geodetic reference datum. The reference datum adopted in India was the Everest ellipsoid that has been developed after adjustments of the Indian geodetic network existing in 1880. Kalyanpur was taken as the initial point (Latitude: 24°07'11".26; Longitude: 77°39'17".57) Hence there were limitations regarding the availability of data and computational methods. Meanwhile a lot of geodetic data was collected and advanced computational methods were available. Hence, it became possible to take up integrated adjustments leading to redefinition of the datum. The Indian geodetic datum or the Everest ellipsoid was based on Great Trigonometrical network which was initiated in April 1802. Further, geoid based on the mean sea level was used as datum for finding out height of any point in the country. SOI continued to have the primary responsibility in this regard. The first level net and the vertical datum was established in 1909. In order to further refine the datum, a second level net was mooted in 1977.

The South Asian region was earlier divided into five parts: four quadrilaterals (NW, NE, SE and SW) and the southern trigon for the adjustment. Owing to the computational limitations, these quadrilaterals could not be adjusted. In 1937-38, an unsuccessful attempt was made to readjust the network. As a result, Everest Ellipsoid was continued to be used as the geodetic datum for India and neighbouring countries. The Indian topographical maps and all geodetic coordinates are based on this datum. However, serious attempts were made to overcome this problem and to redefine the Indian geodetic datum. This will certainly improve the accuracy of the datum and thereby higher accuracies in geodetic coordinates, distances, azimuths and gravity anomalies would become possible.

Contributions of Indian Surveyors

George Everest was most impressed by two Indians, namely Syed Mir Mohsin Husain and Radhanath Sickdhar. Sickdhar who joined him in 1840 as computer, turned out as a mathematical genius and first Indian to be employed in GTS work. He became the right Arm of Everest and besides computations was allowed to make even observations. He computed the height of Mount Everest, which was declared as the highest point in the world. He compiled the first edition of *Auxiliary Tables of the GTS* in 1850. He trained many Indian computers who were employed to complete this great work.

Syed Mir Mohsin Husain, whom Everest picked up from a jeweller shop in Madras in 1824, rose to become his leading instrument repairer. He was born at Arcot and was connected with the family of the *Nawabs*. He had already learnt to take astronomical observations. After 1830 Everest found him particularly useful in petty repairs and adjustments of new instruments besides repairing the old ones. He was appointed in 1824 as instrument repairer in the Surveyor General's Office (SGO) in Calcutta. Subsequently, he was promoted as sub-assistant and in 1843 as Chief Mathematical Instrument Maker. Earlier, in 1836 Everest reported him as:

> "Peculiarly remarkable for his inventive talent, the facility with which he comprehends all mechanical arrangements, and the readiness with which he enters into all the new ideas of others. He later said that without his valuable aid rendered to him by Mohsin Khan, it would have been utterly out of my power to carry into effect my various projects for the remodelment of the instruments, completion of the apparatus for comparing the chains, standard bars, remodeling of 18-inch theodolite and for having been able to introduce my reverberatory lamps into practical use."

His crowning triumph was his successful division of the horizontal circles of two astronomical instruments in 1839 which was considered even better than the original. Mohsin, formerly designated as native artist, a person of natural genius and speed, and a practical turner and mechanic. Further, there were other Indian surveyors and subordinates who had contributed to such a gigantic operation. One was Kusiali, a young man was the head smith of the G.T.S. and helped in the establishment of artificers. Similarly, there was one Ramdheen, head carpenter in the SGO who was trained in the workshop to be an able turner and workman in brass and iron. George Everest developed full faith in Indians and highly commended their hard work, dedication and devotion to duty even under extreme hard climatic conditions and working conscientiously even without supervision.

Geodetic Observations in India

As mentioned earlier, India was pioneer in taking up geodetic observations which had no immediate benefits to the East India Company and later to the Crown administration. Nevertheless, the benefits were for the whole scientific community in the world for understanding the shape of the earth and for having an integrated approach to comprehend the earth system sciences. It included gravitational forces, magnetic forces, levelled heights, tidal predictions to determine sea level, and astronomical observations to determine the latitude and longitude. Observations were made for suitable corrections applied to get the best fitting shape of the geoid. Geodetic triangulation was carried out in India from 1802 to 1878 and was adjusted in 1880 to the Everest spheroid. Inherent weaknesses were discovered by J.D. Graff Hunter in 1916 and by Capt. G. Bomford in the following year. Certain series were considered as secondary and were later taken up for upgradation and densification. It may be mentioned here that in all these initiatives Indian surveyors were involved in the field operations (Appendix II).

Gravitational Forces

In the study of the figure of the earth, the effect of Himalayan masses to the plumb line was felt by Everest himself when he obtained pendulums of Kater's inversible type from England. Capt. A. Pratt in 1852 calculated the actual amount of attraction of the Himalayan masses and of deflection or deviation of the plum line at three stations and thus propounded the *Theory of Isostacy* which was a significant contribution of India to geophysical sciences in the world. Though the actual pendulum observations were initiated in 1865 by Capt. Basevi but redefinition of Everest datum was felt more since 1927 when international spheroid was introduced and used for scientific purposes. Since then, the demand of gravity and magnetic observations were increased and the whole country was covered with a mesh of gravity stations, nearly 15 km apart. It provides valuable information regarding various gravity anomalies, gravity deflections and undulations. For these observations, many absolute gravity stations were established in India.

The contributions of B.L. Gulatee were considered important amongst the scientific community. He had determined gravimetric geoid in the high

Himalayas and also corrected the height of Mount Everest. Further, he also made significant contributions in the geophysical studies.

Magnetic Observations

Magnetic surveys were started in 1840 by Capt. Boilean and the first determination of dip, declination and horizontal force were made in1901 with the help of instruments like dip circle and conventional magnetometer. The magnetic observatories were also established in various parts of the country during 1902-1904. Alibagh (Colaba, Mumbai) and the value of base line of five magnetic observatories were determined in 1920. Modern instruments were used from 1940 onwards with the advent of QHM (Quartz Horizontal-force Magnetometer), BMZ (Magnctomecric Zero Balance), vector magnetometers. Sabhawala Observatory, near Dehra Dun, was commissioned in 1964 and has been continuously recording observations with latest instruments installed there. Magnetic instruments have been calibrated here and are standardized at Alibagh observatory. There are about 182 repeat stations spread over the entire country on which observations are made at an interval of five years for the preparation of epoch charts.

Levelling Operations

Col. Waugh initiated the plan of connecting trigonometric heights to the levelled heights in 1856. The trigonometric heights had so far been determined by reciprocal observations of the vertical angles. Heights were carried over long distances fairly accurately provided the observations were taken at the time of minimum refraction but since the Indian triangulation was large, it was considered to check heights by levelling. The series was commenced in 1858 and by 1864 Karachi was connected with Calcutta - a level line of 2,200 miles. In levelling, both Indian surveyors and the British officers were partners. It was a team effort. In many other series Indians were employed as second observers for simultaneous double levelling.

Astronomical Observations

Astronomical observations of azimuth for the control of direction were made at the start and close of every series of triangles from Lambton's time.

Everest introduced modifications for azimuth observations with circumpolar stars. These observations played a significant role in the determination of astrogeodetic geoid and to improve upon the triangulation network based on the Laplace azimuth. Further, the faith in Indian surveyors was appreciated when by several careful observations, the latitude of a place at Ujjain was carefully worked out from the astronomical observatory set up by Raja Jai Singh Sawai at Ujjain. Similar observatories were also found at Jaipur, Delhi, Mathura and Varanasi. The employment of Indians thus came to the fore, especially training of technical persons needed for survey work in different fields and in particular precision geodetic survey work.

Tidal Predictions and Datum

Systematic tidal observations were initiated in 1876 by installing self-registering gauges all along the coast when difference of three feet was noticed between Bombay and Madras sea levels. A number of Indians were employed for recording and setting up tidal installations, inspecting the observations and training other Indian clerks.

Provision of Ground Control points

Earlier for control work, conventional technologies had been used, such as astro-observation for azimuth; base measurements to check scale distortions; levelling for vertical control; and triangulation, traverse, and trilateration for horizontal control. Further, the first adjustment of Indian geodetic triangulation was completed in 1880. This operation involved a total of 68 series consisting of about 2,700 stations, and ten bases quadrilateral. The whole network, as mentioned earlier, was divided into five parts, *viz.* North West quadrilateral, North East quadrilateral, South East quadrilateral, South West quadrilateral, and Southern Trigon. The detailed field survey included plane table surveys and photogrammetrically plotted sections with field verifications. Next step was fair drawing consisting of mosaicking of field plane table sections, conventional fair drawing and scribing technique. The final step was the printing of maps based on fair drawing originals on plates developed with photographic methods; and later from scribed layers circumventing some photographic processes.

International Terrestrial Reference Frame (ITRF) is a realization of International Terrestrial Reference System (ITRS), which has been a definition of geocentric system adopted and maintained by *International Earth Rotation and Reference Systems Service* (IERS). IERS was established in 1987 on the basis of the resolutions by the International Union of Geodesy and Geophysics (IUGG) and International Association of Geodesy (IAG). The origin of the system was the centre of mass of the earth. The unit of length has been metre. The orientation of the axes was established as consistent with that of IERS's predecessor, Bureau International de l'Heure or BIH in 1984. The Z-axis is the line from the earth's centre of mass through the Conventional International Origin (CIO). Between 1900 and 1905, the mean position of the earth's rotational pole was designated as the Conventional Terrestrial Pole (CTP). The X-axis is the line from the centre through the intersection of the zero meridian with the equator. The Y-axis is the line from the centre to equator and perpendicular to X axis to make a right-handed system. ITRF has been a set of points with their 3-dimensional Cartesian coordinates and velocities, which contributed to an ideal reference system, *i.e.*, ITRF. Each ITRF is identified by the digits of the year as ITRF (year) *e.g.*, ITRF97, ITRF2000 etc. The latest was then ITRF2005. For realization the different techniques are used:

(a) Very Long Baseline Interferometry (VLBI),

(b) Lunar Laser Ranging (LLR),

(c) Satellite Laser Ranging (SLR), Global Positioning System (GPS), and

(d) Doppler Ranging Integrated on Satellite (DORIS).

Each has strengths and weaknesses. Their combination produced a strong multi-purpose Terrestrial Reference Frame (TRF). ITRF has been the best geodetic reference frame currently available. The GRS80 ellipsoid is recommended by IERS to transform cartesian coordinates to latitude and longitude.

Densification of Geocentric Geodetic Datum

The number of stations directly realized by the geocentric geodetic system is not enough for practical use. Worldwide there are several hundred realized stations in ITRF and about twenty stations in WGS84. Therefore it was felt necessary to increase the density of the local stations based on the given stations in each nation or area. Examples of densification projects are NAD83 in North America, SIRGAS in South America and ETRF in Europe.

International Scenario

Most of the international community is switching over to the geocentric reference system. The GRS67, GRS80, WGS84 and ITRS have been the results of the efforts to arrive at the best possible geocentric system. The developed countries have already defined their datum on geocentric system. Some of the developing countries are on the way of realizing their respective datums. At the XX General Assembly of the International Union of Geodesists and Geophysicists (IUGG) in 1992, the International Association of Geodesy (IAG) made the following recommendations in their resolution (IAG,1992) :

(a) That groups making highly accurate geodetic, geodynamic or ocean-ographic analysis should either use the ITRF directly or carefully tie their own systems to it.

(b) That for high accuracy in continental areas, a system moving with a rigid [tectonic] plate may be used to eliminate unnecessary velocities provided it coincides exactly with the ITRS at a specific epoch.

It was apparent that these recommendations allowed for the use of other systems providing they were carefully tied to the ITRS. It was also to be noted that in ITRF the motions of the individual tectonic plates will cause horizontal coordinates to constantly change in time. It was therefore neces-sary to specify which epoch ITRF coordinates refer to and to account for tectonic motion while changing epochs. The countries defining their own datum connect it with ITRF had to comply with the above recommenda-tions of IAG.

Establishment of Indian Geodetic Datum 2007 (IGD-2008)

Strategy

The ideal solution for the present scenario may be about the establishment of the datum based on dense network of permanently operating GPS stations. This provided the users an easy and quick access to the control. Developed countries like US and Canada had thousands of such stations in the form of Continuously Operating Reference Stations (CORS) and Computer Assisted Coding System (CACS) respectively. However, for a developing country like India, it may not be practicable to establish thousands of permanent GPS stations with facilities to transmit the differential corrections. It requires huge resources in order to establish and maintain such a system. The best option for India was to have a balanced approach using available permanent GPS stations along with dense network of campaign mode stations.

Observation Techniques

In India, the only space-based technique available at present is Global Positioning System. Hence, the establishment of Indian Geodetic Datum or IGD2008 was to be based on the GPS technique. In future, if VLBI (Very Long Baseline Interferometry) stations were to set up in the country, they were supposed to use as the highest accuracy points along with the GPS stations to connect the Indian datum with the ITRF as other developed countries have also done in the past.

Data Sources

Realization of a geodetic datum required setting up of an accurate national network of control points. It should include all the continuously operating GPS stations and the campaign mode ground control points. India then had the data sources in the form of permanent GPS stations, GPS stations at tidal observatories and campaign mode stations of GCP Library.

Permanent GPS Stations

India has a dedicated network of 41 permanent GPS stations out of which five are maintained by SOI and rest by other government and autonomous organizations such as Wadia Institute of Himalayan Geology (WIHG), Geological Survey of India (GSI), Indian Institute of Geomagnetism (IIG), Indian Institute of Astrophysics (IIA) and several other organizations. These permanent GPS stations had been established to monitor the geo-tectonic activities in the country as a long-term programme funded by the Union Department of Science and Technology. GPS data from all these stations is being received in the National GPS Data Centre located in the G&RB in Dehra Dun (Fig 4.2). All these stations were to be used for realization of the datum and contribute towards establishing the Continuously Operating Reference Stations or CORS in the country.

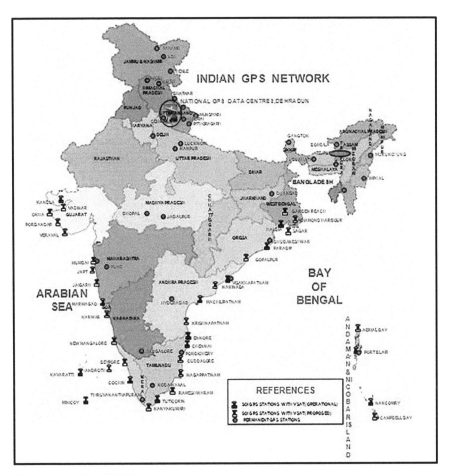

Figure 4.2 *Indian GPS Network*

Aizwal	Dhanbad	Kothi	Pondicherry
Almora	Gulmarg	Kolhapur	Port Blair
Anini	Gangtok	Leh	Pune
Bhatwari	Guwahati	Lucknow	Rajkot
Bhopal	Hanle	Mumbai	Shillong
Bhubaneshwar	Imphal	Munsiari	Tezpur
Bomdilla	Jabalpur	Naddi	Trivendrum
Chitrakut	Jaipur	Nagpur	Tirunelveli
Dehra Dun (SOI)	Kanpur	Panamik	Visakhapatnam
Dehra Dun (WIHG)	Kodaikanal	Pithoragarh	Lumami
Delhi			

Table 4.1 *Continuously Operating Reference Stations in India*

GPS Stations at Tidal Observatories

In addition to the permanent GPS stations, there were 14 GPS stations maintained by SOI at tidal observatories which have been established as part of modernization of Indian tide gauge network (Table 4.2). These tidal observatories were equipped with real time transmission facilities through dedicated VSAT network. GPS data is being transmitted in real time to National GPS Data Centre. Few more GPS receivers were to be added to this network soon (Fig 4.3).

Kandla	Kawaratti	Machhilipatanam	Port Blair
Marmagoa	Minicoy	Vishakhapatanam	Nancowry
Cochin	Tuticorin	Paradeep	
Chennai	Ennore	Haldia	

Table 4.2 *GPS and tidal stations*

Figure 4.3 *Tide gauge network and 5.2m VSAT Antenna at Geodetic & Research Branch, Dehra Dun*

Ground Control Points Library

Initiation of the GCP library project had paved way for the realization of Indian horizontal datum or IGD-2008, as it was commenced for setting up of precise and consistent framework of reference points in geocentric co-ordinate system spread across entire country. Setting up of the GCP library was planned to be completed in three phases with following objectives:

- Setting up of precise and consistent GCP Library in geocentric co-ordinate system for the entire country at a spacing of approximately 30 km.

- The GCPs will have permanent monumentation with security fencing and identification boards and was to be made available to the user community for their scientific, engineering and other developmental projects requirements (Fig 4.4).

- It was also proposed to set-up beacons at these stations for transmitting GPS differential corrections so that the GPS users can get their positions within 1 metre in real time.

- The GCPs were to serve as the store house of information such as precise horizontal control, MSL heights, gravity and magnetic values needed for various applications.

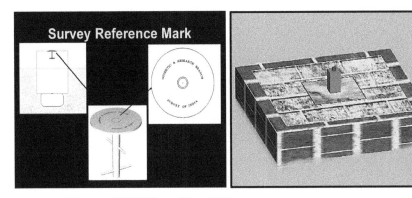

Figure 4.4 *Design of GCP Library Phase-I Monument*

Till 2002, about 3,500 geodetic horizontal control points and about 4,000 plus precision and high precision height control points were established in the country. These control points were although sufficient for extending topographical points for mapping on scale 1:50,000, but considering the requirement of large-scale mapping, there was a need to densify the precise control for getting quality digital data. SOI had planned to use the existing control as reference and densify it by using GPS techniques. Such densified control was to be provided in a fashion to have about 120 points in each map sheet area of 1:250,000 scale topographical sheet (110 km to 110 km) within the next two years' time for entire country.

Phase I

In phase I, nearly 300 *Zero Order* (highest precision) Ground Control Points (GCPs) at spacing of 250-300 km had already been established in geocentric coordinate system for the whole country (Fig 4.5). These Zero Order control points were defined the Indian horizontal geodetic datum. GPS observations were to be carried out for 72 hours at these stations and their locations were independently computed using 3 to 4 IGS (International GPS Service) stations which defined the International Geocentric Reference Frame ITRF – YY, where 'YY' represent the epoch. The latitude and the longitude of the *Zero Order* control points were computed using scientific GPS software, Bernese, to an accuracy of milliarc second amounting to 3 cm on ground. In addition to having precise horizontal coordinates (latitude and longitude), they were also provided with precise heights above mean sea level using levelling.

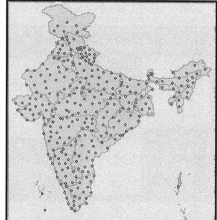

Figure 4.5 *GPS Network*

Data Processing and Adjustment

As per the IAG recommendations, the datum was to be connected with ITRF. India has two IGS stations, one at Bangalore (IISC) and another at Hyderabad (HYDE). In addition to these, the IGS stations were available around South Asia in Beijing (BJFS), Ankara (ANKR), Wuhan (WUHN), Lhasa (LHAS), Poligan/Bishkek (POL2) and Kitab (KIT3) and were also be used as fiducial points. The Indian permanent GPS stations were planned to be connected with these IGS stations to get the datum in terms of ITRF.

The GCP Library Phase-I stations were observed independently, and the data had been processed with respect to IGS stations to determine the coordinates in terms of ITRF. These coordinates can be used for general mapping purposes until the datum is realized. The network adjustment was to be carried out.

As already mentioned above, the GCP Library phase-I stations were observed independently and not in well-defined network mode. For adjustment, a network was required to be formed. When the CORS were to be included in the network, hundreds of baselines were to be formed by the common observations among CORS and Phase-I GCPs. By carefully selecting few hundred baselines, a network of more than 300 interconnected triangles were formed. The network was in such a way that all the GCPs and most of the CORS were connected.

In the first step, all the baselines were computed using Bernese GPS Data processing software. The second step was to be connecting the CORS to the IGS stations. This was necessary to connect the Indian datum with the ITRF as per the IAG recommendations. The CORS network was to be adjusted and coordinates were to be computed on the epoch 1st January 2007, taking the IGS stations as fiducial stations. This was to be done using scientific GPS data processing software (Bernese 5.0). The CORS was then to be used as known stations and the GCP Phase-I network was to be adjusted. The slope distances were to be computed using this software and was to be reduced to ellipsoidal arc distances using a computer programme. These ellipsoidal arc distances were to be taken as observables and the adjustment were carried out using ADJUST programme developed by National Geodetic Survey (NGS), USA and freely available on the net. This realization was to be based on ITRF2005 which had the epoch 1st January 2000. Using the velocity model, the epoch of this realization was to be kept as 1st January 2007.

Phase II

In this phase, additional 2,200 GCPs of first order were to be established at an average of 30 km spacing all over the country. GPS observations were to be carried out for 6-8 hours at these stations and the base lines were to be computed using adjustment software. The station coordinates were supposed to be adjusted with zero order control points as per usual survey principle *i.e., whole to part*. The expected accuracy of horizontal coordinate was in the order of 0``.002 amounting to 6 cm on ground. The first order control points were to be connected by levelling. Work was then under progress. It was to be completed in 12 months, if top priority was given to this job and some activities and resources were outsourced and extra equipment were procured on urgent basis.

Phase III

In phase III (image control points) each sets for providing GCPs were to be avoided. GCP observations and adjustments were to be a part of integrated network of zero and first order GPS networks of triangles. Location of GCPs can only be decided by marking the first on imageries after their availability. This was a huge task and was considered to be a fit case for outsourcing. This phase III job can be started after the availability of imageries and phase II GCPs whichever was later. It was then estimated that by employing 80 teams along with vehicle, the job can be completed in a year. By outsourcing some of the tasks can be completed even in nine months.

Establishing GPS Data Centre

There was an attempt to have a National GPS Data Centre for earthquake monitoring in the GBO Compound (17 East Canal Road) of SOI in Dehra Dun. It started functioning with data received from the three permanent stations situated in Pune, Thiruvananthapuram and Bhubaneshwar. Further, for storage, dissemination and management of huge GPS data, a national data centre was to be established. It was also proposed to have a secondary data centre later in the Survey Training Institute, Hyderabad. Similar data centres were set up in the United States, Japan, Switzerland and other countries of Europe. Furthermore, tracking stations were proposed as well. These stations were to be equipped with precision, dual frequency, P-code receivers operating at thirty seconds sampling rates. The IGS or International

GNSS service supported nearly 150 globally distributed stations which were continuously tracking and were accessible through phone lines, network, satellite connections hence allowing quick and automated downloading of data on a daily basis. It was proposed to have about 18 permanent stations in the first phase, including five stations in the peninsular shield. All these centres were to supply information to the National GPS Data Centre in Dehra Dun.

Earlier Initiatives

G&RB has been doing the geodetic and geophysical measurements for entire country in connection with national mapping and for other geodetic and geophysical research. Geodetic and geophysical measurements were also take-up in the areas suspected to have either the seismic activities or local deformation due to existence of fault or thrust lines. The importance of monitoring precursor of earthquakes was realized and a systematic data collection for the purpose of crustal movement studies were initiated in 1972. Geodetic and geophysical data collected in the past was being utilized for the purpose whenever the observations were repeated.

India had experienced several great earthquakes during the past *viz.* 1897,1905, 1934 and 1950 in Himalaya and similar events elsewhere such as in 1991 Uttarkashi in 1991, Latur in 1993, Chamoli in 1999 and Gujarat in 2001. Further, SOI had taken up geodetic and geophysical measurements in the Gujarat region immediately after Bhuj earthquake which occurred on 26 January 2001. GPS, gravity, geomagnetic and levelling observations were planned during February 2001 to assess the magnitude of surface and sub-surface (geological) deformation in post-earthquake scenario (Fig 4.6).

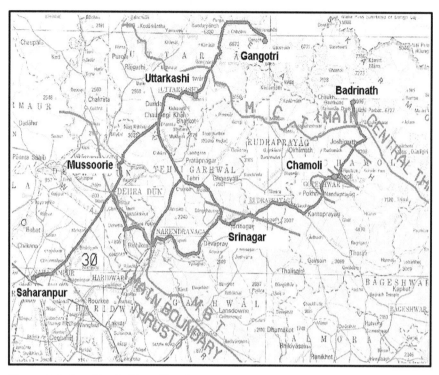

Figure 4. 6 *Levelling in Uttarakhand*

GPS observations were carried out on twenty-eight old existing stations covering entire Gujarat state. The observation period was kept at 24 hours in order to get more precise results. Data collected during the fieldwork was processed using precise ephemerides and with Bernese scientific software. GPS derived coordinates for all stations were compared against the existing values and differences in their horizontal positions were taken into consideration for forming the opinion about the magnitude of surface deformation arising out of the earthquake. Though, the available data was quite old, and accuracy of their measurements cannot be taken at par with better instrumentation, yet estimation of shifts in horizontal positions provided some information about the crustal deformation. Based upon these estimates, changes of varying degrees were detected in eastern or northern directions. Repeat gravity and geomagnetic observations indicated about the geological changes. Earlier, SOI had established a network of gravity and geomagnetic stations in the area which were re-visited to assess the magnitude of variations in available values.

Existing data pertained to the year 1983-84 gravity profile along Ahmedabad-Rajkot section which recorded a decrease in gravity ranging from 38 to 328 micro gal. Along the Rajkot-Porbandar-Okha-Jodiya section, there was a mixed response to gravity changes with significant increase at Ribeda village (south of Rajkot), Porbandar, along Tupini-Jodiya and significant decrease along Lamba-Okha segment. Along Jodiya-Bachau-Mundra-Bhuj Lakhpat, there was an increase in gravity values along Bhachau-Mundra section. The standard gravity station at Bhuj airport did not show any significant variation in existing 'g' value. Increase in 'g' value was more pronounced around Lakhpat.

Geomagnetic observations at six repeat stations were taken after the earthquake and recorded values of horizontal and vertical components were compared against the values observed during 1995-96 and 1999-2000. No significant magnetic variations observed except at Bhuj. Vertical intensity at this place showed increase by 261 nT over 1999-2000 values, which was quite prominent. This large variation reflected the change in the composition of rocks beneath the earth's crust. Further, there were some vertical movements at few places where the levelling observations were carried out during February 2001, but the results could only be validated after completing the entire level net in the state. However, the results provide only a broad indication of surface and geological changes caused by the earthquake. Repeat GPS observations at suitable interval were recommended in order to arrive to some conclusion about the horizontal movements in the area. There was a need to establish a dense network of level lines and gravity stations in the area and repeat the existing level and gravity stations.

Monitoring Crustal Deformation & Fault Displacement in Kachchh Region

G&RB has also been involved in geodetic and geophysical measurements in the areas suspected to have either seismic activities of local deformation due to fault and thrust lines. The importance of monitoring precursor of earthquake was realized and systematic data collection for the purpose of crustal movement studies started in 1972. Data collected in the past was being utilized and observations were repeated. SOI took up crustal movement studies in earthquake prone areas in Gujarat region immediately after Bhuj earthquake which occurred on 26th January 2001. GPS, gravity, geomagnetic and levelling observations were taken immediately after the event to assess the magnitude of surface and sub-surface (geological) deformation in post-earthquake scenario.

GPS observed baselines could yield sub-centimetre accuracies. The comparison was made between conventional techniques, *i.e.,* GT triangulation having limitation of 10ppm accuracies, and GPS techniques which were much accurate. The broad indication of movement of the crust have been noticed (Fig 4.7). Further, repeated gravity and levelling observations provided an important clue about vertical movement of the crust caused by seismotectonic activities in the concerned region. The amount of gravity variations at each station were not directly related to the upliftment or even subsidence. However, the changes in the gravity values can partly be correlated with vertical deformation. Furthermore, it was recommended that gravity and levelling observations should be carried out simultaneously during the next attempt in order to get better results. Geomagnetic results at Bhuj station had shown entirely different trends from rest of the stations. Hence, it was also recommended that there should be dense network of geomagnetic stations around Bhuj.

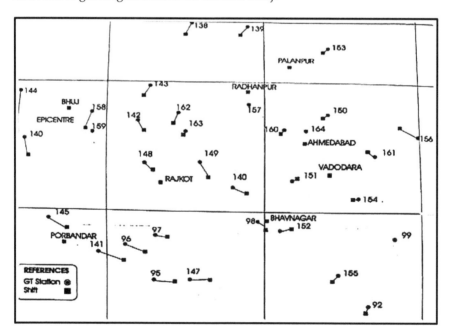

Figure 4.7 *Crustal deformation in Kuchchh region*

Analysis of Impact of Tsunami in Andaman & Nicobar Islands

The Indian Union consisted of thirty-two states and union territories including groups of islands in the Arabian Sea and Bay of Bengal. The latter groups of landmass were known as Andaman and Nicobar Islands covering an area of about 850 sq. km which is comparable to several smaller nation-states of the world. They represent the elevated portions of submarine mountains and on an alignment where the Indian plate meets the Burma plate. Further, there are three islands in the Andaman group : North, Middle and the South. Mount Harriet (460m) in South Andaman Island has been the highest point. Little Andaman lies further to the south, separated by the Duncan passage from the main group of islands. The Great Nicobar is the largest of another group of nineteen islands. There are two other islands outside the two main groups: (b) Barren island, and (b) Norcodam islands. The former island has been known for its volcanic activities.

SOI had a long association with Andaman and Nicobar islands. The entire mapping of these islands had been done by this organisation based on the survey control provided by astronomical observations during 1861-87. Hence the locations of all the islands were observed separately. The maps were prepared considering the islands as separate entity. Lately, with the introduction of the GPS, efforts were made to connect these islands among themselves as well as with the mainland. This technology was used during 1994-95 with the active help of National Hydrographic Office, Ministry of Defence.

In 1861, the latitudes were based on astronomical observatory in Chatham islands which were 11°41'13.00 +/-0.°11N. Similarly, longitude was calcu-lated as 92°41'44".00 +/-0°.30E. Height was based on Ross island benchmark (near Port Blair) which was 7.77 feet above the meal sea level. Further, the triangulation in these islands was based on the following values at Camorta Observatory:

~Latitude: 8°02'20".79

~Longitude : 93°31'55".05

~Height above msl: 6 feet (benchmark built near jetty).

In 1994-95, a total of 28 stations were identified for observation. Repeat observations were carried out on all these pre-existing GPS stations in 2003-04. The detail analysis of the results obtained from the two epochs of observations, *i.e.*, 1994-95 and 2003-04, revealed that there has been a horizontal displacement of 2 to 6m shift in position of most of the points in the islands in last ten years interval. India has permanent GPS stations in Dehra Dun, Shillong and Thiruvananthapuram. In Andaman and Nicobar islands, GPS were used not only for mapping but also for monitoring and assessing changes occurred in the landmass due to massive earthquake and consequent tsunami hitting these islands on 26th December 2004. Survey of India team was again deployed to quantify the shift in the position of islands by carrying out observations on the pre-existing stations in post tsunami scenario.

In order to assess the changes in the landmass, several sets of observations were taken with the help of GPS. Differences in heights were also calculated. Following sets of observations have been taken :

(1) Comparison of pre and post tsunami vectors between mainland and islands

(2) Comparison of pre and post tsunami vectors from mainland to islands, processed with precise ephemerides with respect to permanent stations

(3) Comparison of coordinates between pre and post tsunami observations, processed with broadcast ephemerides with respect to GT stations

(4) Comparison of coordinates between pre and post tsunami observations, processed with precise ephemerides with respect to permanent stations

(5) Pre and post tsunami ellipsoidal (WGS 84) heights of island stations, processed with precise ephemerides with respect to permanent stations.

As a first step of the study, the three island stations at Aerial Bay, Campbell Bay and Port Blair were measured with respect to two mainland stations at Nanmangalam, Chennai and Yarda, Visakhapatnam. The maximum difference of 3.7988m was found in the Nanmangalam-Aerial Bay vector. However, in all the six measurements, the minimum difference was about 1.41m. In a next step, the pre and post tsunami vectors were calculated with precise ephemerides and with permanent stations. The maximum difference was found in Aerial Bay-Post Blair vector of -1.8284m. The minimum was along Aerial Bay-East Island. The results indicate that there was no pattern of movement or changes in the islands (Fig 4.8).

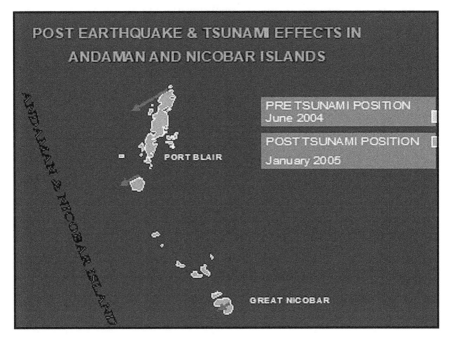

Figure 4.8 *Pre and post tsunami positions*

In order to further assess the changes, the pre and post tsunami coordinates of the remaining 12 stations in the islands were calculated with broadcast ephemerides and with respect to GT stations. The maximum change was in Passage island (-0.124") and in Meroe (-0.198"). Here also there was no pattern *per se*. Further, same exercise was carried out with precise ephemerides and with respect to permanent stations. The maximum change in latitude and longitude was in Tarassa (-0.102" and -0.195"). If we compare the results, we will find that, based on the observations at GT stations, changes vary from island to island and as such there is no pattern. Some islands like Meroe, Tasassa, Aerial Bay and East Island had undergone more

changes in comparison to other islands. The pattern of shift in position of
the three island stations was calculated (Fig 4.9). Shift in position of islands
with respect to each other was also evident from the baseline measurements
of Port Blair-Udaygarh and Port Blair-Long Island. The data of pre and
post tsunami changes in heights on WGS 84 ellipsoid was processed with
precise ephemerides and with respect to permanent stations. Here again
the maximum difference was found in Meroe (-2.14m) and Tarassa (-2.86m).
It may be mentioned here that during the visits of Kurz Sulpiz and R.D.
Oldham of the Geological Survey of India in 1866 and 1884 respectively
considered these islands as residue above water of submarine subsidence.
Further, M.V. Portman, officer-in-charge of Andaman (1879-1900), also
believed subsidence in immediate past. Hence, the change in heights was
not a new phenomenon.

Shift in Position

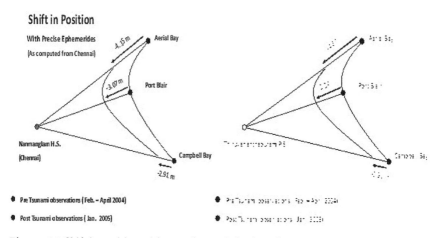

Figure 4.9 *Shift in position with regard to mainland stations*

Preliminary investigations of the GPS results from April 2004 and January
2005 epochs of observations reveal that there was a small change in position of Port Blair station. GPS observations were carried out at this station
immediately after tsunami. A vertical subsidence of approximately 1.2m
was observed. Further, tide pole observations were also carried out at
Chattam jetty by Andaman & Lakshadweep Harbour Works. Preliminary
levelling between benchmark and bed-plate at the Chattam jetty was carried
out by SOI as well. Tide pole observations were reduced to chart datum
and comparing the observed values with the predicted tides indicated a
subsidence of 0.80m at Port Blair. Further, the difference in heights between
the tidal benchmark and the bed plate at Chattam Tidal Observatory was

computed by running a levelling line after tsunami. These observations indicated a subsidence of 0.40m. Thereby a total subsidence of 1.20m has been observed. The deformation study was analysed with respect to the stations in the mainland.

There had been a history of changes of landmass in all the three directions in Andaman and Nicobar Islands as per the records now available. Hence, the changes brought by tsunami in December 2004 are not that alarming. However, repeat GPS were observations in campaign mode have to be carried out and the results will have to be analysed to detect the phenomena of surface deformation. Further, modernization of tide gauges is required in order to study the sea level changes. Such changes are now considered to be precursor for earthquakes followed by tsunamis. Preferably, the tide gauges should be equipped with GPS. Further, there has been post-tsunami changes in the landmass of Andaman & Nicobar islands as well (Fig 4.10).

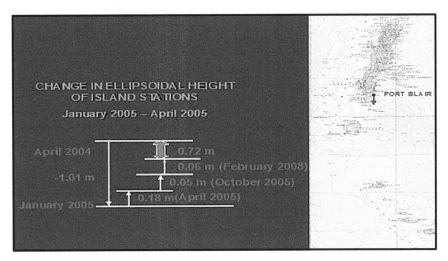

Figure 4.10 *Post-tsunami changes in height*

Nevertheless, the changes brought by the tsunami may a case for making a case for fresh topographical mapping because there have been changes in the mean sea level; there has been subsidence and emergence at different places; coordinate and control points have changed; and values of baseline have also changed. In addition, the shocks were still frequent in the concerned area. Hence a right time to launch this exercise has also to be decided.

Recent Initiatives

During the Global Spatial Data Infrastructure (GSDI) conference in Bangalore in 2004, emphasis was also on the development of geodetic framework
for the Asia and Pacific region. Of course, such framework was to be a part
of the global initiative. There was a separate committee of GSDI looking
into this matter. As mentioned earlier, most accurate global terrestrial reference frame has been the ITRF, maintained by the *International Earth Rotation
and Reference Systems Service* (IERS). This frame has increasingly been used
as the backbone for the national reference frames. The whole initiative has
been voluntary and depends solely on commitments of national authorities
and their specialized agencies. Such cooperation is likely to develop inter-
dependence and provide global services. Consequently, national geodetic
infrastructure cannot be developed in isolation. Hence, it is a part of global
geodesy enterprise.

Since, more and more issues are becoming common, such as for economic,
infrastructure and sustainable development, there has been a need for coordinated and comprehensive monitoring of the earth system. As a result,
international consensus has been developing where a number of countries
are participating in *Global Earth Observation System of Systems* (GEOSS). It
aims to contribute for societal benefits such as disaster management, climate
and climate change, land and water resources. Their management directly
depend on sound geodetic framework and observations. Further, United
Nations Committee of Experts on *Global Geospatial Information Management* or
UN-GGIM has also acknowledged the increasing need of more precise positioning services thereby pleaded for global cooperation in geodesy. In July
2013, it was resolved to develop a global geodetic reference frame (GGRF).

The working group on the GGRF was created in January 2014. Working
Group meetings have been held and draft resolution was developed. It
was endorsed by United Nations Economic and Social Council (ECOSOC)
in November 2014. ECOSOC presented the proposal to the UN General
Assembly in early 2015 for ratification. In February 2015 the UN General
Assembly adopted the resolution in favour of "A Global Geodetic Reference
Frame for Sustainable Development" – the first resolution recognizing the
importance of a globally-coordinated approach for geodesy. Recognizing
the growing importance of the subject in societal benefits, the members of

subgroup on National Geodetic Infrastructure had deliberated upon the issues related to requirements of geodetic infrastructure in India and put their views in the form of a report. The focus of a report was on status of the current geodetic infrastructure in India and possible requirements for future.

Further, the report provided a comprehensive overview of requirements for geodetic observations, products, and services on national as well as global levels. Based on these initiatives, general considerations for the design of the global and national geodetic observing system were mooted. Thus, the material presented in the report was not only applicable to the specific situation in India but rather constitutes a broad basis for the design of national and global geodetic infrastructure. In many countries, the national geodetic authority has been involved in or associated with time and frequency transfer, and geodetic infrastructure has been used for these purposes. However, this report does not address the application of space-geodetic infrastructure for time and frequency transfer. The basic measurements in space geodesy have been time measurements. Moreover, the geodetic infrastructure in India has not been used for time and frequency applications, and, therefore, this aspect was not considered in the scope of this report.

GTS to GPS

From 9-12 February 2003, a national seminar was organised on *GTS to GPS: Geodesy on the Move* in Dehra Dun, the headquarters of SOI (Fig 4.11). This event was meant to observe the two hundred years of the Great Trigonometrical Survey. It was attended a wide range of participants from India and abroad. The seminar provided an opportunity to the geo-scientific community to contemplate not only on our past traditional techniques and achievements, but also a forum to discuss the challenges of the 21st century, emerging technologies and how they can be put to effective use in order to achieve the national objectives. The deliberations were based on the following themes:

(a) Cadastral Surveying and LIS: Rural and Urban Scenario.

(b) Surveying & Mapping for National Defence.

(c) Geodetic Datums: Horizontal & Vertical.

(d) Space Based Technologies for Surveying.

(e) Geodetic & Geophysical Measurements for Earthquake Monitoring.

(f) Marine Geodesy.

(g) Spatial Data Infrastructure: Issues, Challenges & History of GTS.

(h) Natural resources Management.

(i) Microzonation & Disaster Management.

Figure 4.11 *GTS to GPS: Geodesy on the Move*

Some of the themes were conventional. Nevertheless, new themes and technologies were also included. Earlier, the work done by the GTS were handed over to Geodetic Branch, which is now known G&RB. It still carries out conventional triangulation surveys. However, with the passage of time and advent of modern technologies, the activities of this directorate have diversified. Keeping in pace with advanced countries, it has been engaged in a number of research and development projects like providing control points including CORS points, surveys for hydroelectrical projects, building tunnels, heavy industries, observing deflections in the dam structures and tilts in the high minarets and survey of funnel areas for airports. Furthermore, it has been actively engaged in the study of glaciers, monitoring earthquakes and structural movements, conducting geophysical studies in Himalayan region and the like. Attempts were also made to compare the results of distances measured between the baselines with old instruments with re-measuring the same distances with GPS and reducing the results to Everest spheroid. In most of the cases GTS observations carried nearly two centuries ago still hold good (Nag & Nagrajan, 2003) .

Chapter 5

Map Policy & Data

The science of surveying and mapping encompasses a broad range of disciplines including cartography, remote sensing, GIS and GPS. The technology of digital image processing, the GPS and GIS are being used to integrate, processed and analyzed the spatial data for sustainable development of the natural resources. The digital data (*e.g.* high-resolution images, scanned aerial photographs, scanned maps and digital orthophotos) are currently available for this purpose. GIS has been a powerful decision-making tool in the hands of cartographers. Beginning with a computerized topographic map as its base, GIS overlays and integrate graphic and textual information from different databases. The end result has been a customized and reliable means that can support decision making, problem solving and provides almost instantaneous answers to complex questions.

Introduction of digital cartography and use of digital cartographic data base for bringing out updated maps was initiated in SOI in early eighties as an inhouse activity. Consequently, complementary developments associated with the use of digital cartographic database such as formulation of map data structure, development of data exchange formats translators etc. were also undertaken using the resources then available with the department. As a leader in this field, expertise of the department was extended to many scientific government departments or universities and today the standards are being adopted by many institutions. The digital data produced by SOI is in use in many government departments and its undertakings after obtaining clearance from Ministry of Defence.

While considering g-readiness of India with special reference to topographic data, it was logical that the issues related to dissemination of high-resolution geographical data in other fields such as hydrology, climatology, geology and seismology, being generated by other government agencies, was also considered for the purpose of maintaining consistency. At the time of integration of data, just the cartographic base, on its own, was insufficient for any development initiative or research work. The objective was to meet the requirements of socio-economic developmental activities, conservation of natural resources, planning for disaster mitigation and infrastructure development and the like. Accordingly, in May 2005, two map series were accepted by the government. Further, in the *Guidelines for Implementing National Map Policy*, the details of the availability of data were announced a year later, *i.e.* in May 2006.

Restriction on Data Dissemination

Restriction policy on map data had been changing from time to time according to the assessment of internal and external threats perceptions of ministries of External Affairs, Defence and Home Affairs. However, in recognition of the need for geographical data in scientific and educational programmes, classified data was being permitted for bonafide use after going through certain procedures. In view of the introduction of digital technology, in the planning process, the requirement of digital data in GIS environment, the Government of India had allowed ten government organizations to digitise the bulk of topographical data from unrestricted topographical maps at 1:50,000 scale. This derestriction was aimed at making digital geographical data available to scientists, planners, educationalists and other resource users.

National Map Policy 2005

All socio-economic developmental activities, conservation of natural resources, planning for disaster mitigation and infrastructure development required high quality geospatial data. The advancements in digital technologies made it possible to use diverse spatial databases in an integrated manner. The responsibility for producing, maintaining, and disseminating the topographic map database of the whole country, which was the basis of all spatial data, vested with the Survey. Then SOI was mandated to take a leadership role in liberalizing access of such data to user groups without jeopardizing national security. To perform this role, the policy on dissemination of maps and spatial data needed to be clearly stated.

Objectives

The then objectives were:

(a) To provide, maintain and allow access and make available the National Topographic Database (NTDB) of the SOI conforming to national standards.

(b) To promote the use of geospatial knowledge and intelligence through partnerships and other mechanisms by all sections of the society and work towards a knowledge-based society.

Two Map Series

To ensure that in the furtherance of this policy, national security objectives were fully safeguarded, it had decided to have two series of maps namely:

(a) *Defence Series Maps (DSMs):* These were the topographical maps (on Everest/WGS-84 Datum and Polyconic/UTM Projection) on various scales (with heights, contours and full content without dilution of accuracy). They were to cater for defense and national security requirements.

This series of maps (in analogue or digital forms) for the entire country was to be classified, as appropriate, and the guidelines regarding their use will be formulated by the Ministry of Defence.

(b) *Open Series Maps (OSMs)*: OSMs were brought out exclusively by SOI, primarily for supporting development activities in the country. OSMs was to bear different map sheet numbers and was to be in UTM Projection on WGS-84 datum. Each of these OSMs (in both hard copy and digital form) was "Unrestricted" after obtaining a one-time clearance of the Ministry of Defence. The content of the OSMs was different. SOI was to ensure that no civil and military *vulnerable areas* and *vulnerable points* (VA's/VP's) were shown on OSMs.

From time-to-time SOI issued detailed guidelines regarding all aspects of the OSMs like procedure for access by user agencies along with dissemination and sharing of OSMs amongst user agencies with or without value additions, ways and means of protecting business and commercial interests of SOI in the data and other incidental matters. Users were allowed to publish maps on hard copy and web with or without GIS database. However, if the international boundary or coastline is depicted on the map, certification by SOI was necessary. In addition, the SOI was by then preparing city maps. These maps were planned to be on large scales on WGS-84 datum and in public domain. The contents of such maps were to be decided by the SOI in consultation with Ministry of Defence.

National Topographic Data Base (NTDB)

SOI was entrusted to create, develop, and maintain the *National Topographical Data Base* (NTDB) in analogue and digital forms consisting of following data sets:

(a) National Spatial Reference Frame,

(b) National Digital Elevation Model,

(c) National Topographical Template,

(d) Administrative Boundaries, and

(e) Toponomy (place names).

Both the DSMs and OSMs were to be derived from the NTDB. The recent *National Geospatial Policy 2022* has also re-emphasized and entrusted this job to SOI for this purpose (vide Chapter 11).

Map Dissemination and Usage

• Open Series Maps of scales larger than 1:1 million either in analogue or digital formats were to be disseminated by SOI by sale or through an agreement to any agency for specific end use. This transaction was to be registered in the *Registration Database* with details of the receiving agency, end use etc.

- Through the agreement, SOI was to allow a user to add value to the maps obtained (either in analogue or digital formats) and prepare their own value-added maps.

- The users were expected to share these maps with others – the information of all such sharing was also required to be logged in the *Map Transaction Registry*.

Applicability of Previous Instructions

The Ministry of Defence from time to time had issued detailed guidelines on various aspects of map access and use. These instructions were to continue to hold good but for the modifications cited herein. Early 2003, SOI took up a major initiative to digitise the whole topographical base of the country at 1: 50,000 scale. Incidentally, this scale had been very popular with the geomorphologists wherever maps were available at this scale. By the end of this year, all the 5,100 sheets were digitised. The digital data allowed to have a lot of options to play around and was to be extremely useful for terrain and geomorphological analysis as well. Due to the earlier restriction policy for providing maps and digital data, an attempt was made to bring out another set of maps at a different datum and projection.

Further, for the socio-economic developmental activities, conservation of natural resources, planning for disaster mitigation and infrastructure development required high quality spatial data. The advancements in digital technologies had made it possible to use diverse spatial databases in an integrated manner. The mapping scales were 1:25,000, 1:50,000 and 1: 250,000. It was to have heights and contours. The whole series covering the entire country was restricted. There were some differences between the existing topographical maps and this one. However, options for WGS-84 and UTM projections have been included. This development will come in the way of geomorphological mapping in India. As indicated earlier, both the *Defence Series of Maps* (DSMs) and *Open Series of Maps* (OSMs) were to be derived from the NTDB (Fig 5.1).

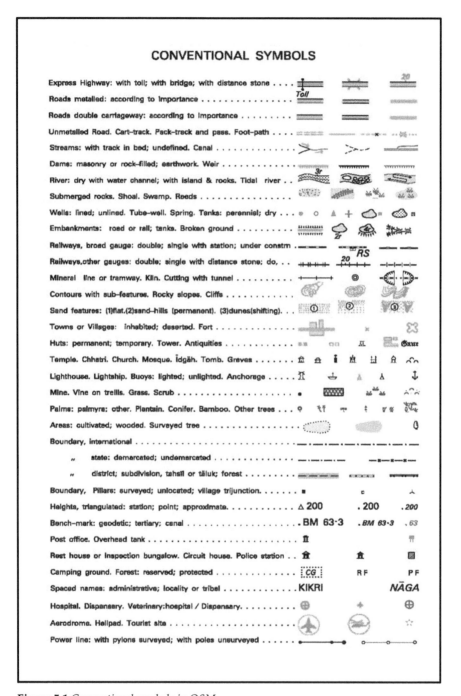

Figure 5.1 *Conventional symbols in OSM*

MAJOR DETAILS	SUB DETAILS
GENERAL	Latitude/Longitude Name of State/District/Administrative index Topographical sheet Number/Year of Survey/Edition/Index to sheets Magnetic variation from true North direction Map reference Bar scale/Representative Factor
ROADS	All Roads
TRACKS	All Tracks, pass, footpath
STREAMS/CANALS	All streams/canals
DAMS	All earthwork dams
RIVERS & RIVERBANKS	All rivers with details, banks, islands
WELLS	All wells/tube wells/springs
WATER FEATURES	All Tanks
EMBANKMENTS	All embankments, Road/rail/tank
RAILWAYS	All gauges with stations, tunnels
OTHER LINES	Light railways or tramway
CONTOURS	Contours with sub features
ICE FORMS	All features
SAND FEATURES	All sand features
TOWNS OR VILLAGES	Inhabited, deserted and forts
SETTLEMENTS	Huts, Towers, Antiquities
RELIGIOUS PLACES	All including tombs/graves
SHIP	Light ship, buoys, anchorages
LAND FEATURES	Mines, Vine on trellis, grass, scrub, cultivation limit
PLANTATIONS, TREES	All features
BOUNDARY, BOUNDARY PILLARS	International to village, Forest, all boundary pillars, village trijunctions
HEIGHTS	Spot, approximate
BENCHMARKS	Geodetic, tertiary, canal
OFFICES	All post/telegraphic/police stations
BUNGALOWS	All including camping ground
FOREST	Reserved Protected
NAMES	Administrative/Locality or tribal

Table 5.1 *Topographical features in the OSM*

The contents which are related to landform or terrain analysis or even applied geomorphological mapping were as follows:

A. Land Based Features

1. Stream: All stream, Canal, Earthwork

2. Riverbanks: Shelving, Steep 3-6m, Over 6m

3. River: All rivers, With islands and rocks, Tidal rivers, Submerged rocks, Shoal, Swamp, Reeds

4. Water features: Spring, Perennial tanks, Dry tanks

5. Embankments: Tank

6. Ice forms: Glacier, Moraine, Crevasses, Scree, Snow

7. Sand feature: Flat, Sand hills & dune, Shifting dune

8. Land feature: Broken ground, Mine, Vine on trellis, Grass, Scrub

B. Cultural and Administrative Features

1. Towns and villages: Inhabited, Deserted, Fort

2. Settlements: Huts, Permanent, Temporary, Tower, Antiquities

3. Roads: Metalled roads according to importance (NH/SH etc), Un-metalled roads according to importance

4. Tracks: Cart track, Pack track and pass, Footpath, and bridge

5. Plantations: Palms, Palmyra, Plantain, Bamboo, Other trees

6. Boundary: International, State, District, Sub-division, Tahsil or Taluk, Forest, Village

Apart from above, the new series was having usual features contained in the topographical sheets, such as names of state, district, sheet number, year of survey, map reference, bar scale, administrative index, railways, post offices, bungalows, and index to sheets.

Advantage of Digital Data

The most important features of the new series were as follows:

(a) Topographical information was to be available in digital form.

(b) Derived maps showing the physical features were to be supplied without any restrictions.

The availability of digital data on WGS-84/UTM series was extremely helpful for further research in terrain analysis and geomorphology as this series had been compatible with (a) remote sensing data then available, and (b) global positioning system (GPS). Hence, temporal changes and precise positioning of different features could now be observed. This was to lead to new thrust areas for geomorphological research where minor change must be monitored very precisely such as bad land formation, river erosion and land slides. The research emphasises shifted to *'process'* - this aspect was not given due importance due to the nature of data then available, *i.e.* topo-graphical sheets or air photos. Process-based studies have been important in India as almost all the geomorphological features do exist in the country in one form or other.

Availability to derived maps showing different physical features had put onus on SOI to prepare such maps which were available without any restrictions. However, for such an initiative, it was imperative to analyse the requirements of different users. Further, in view of the present-day require-ments, SOI planned to provide digital topographical databases for entire country on 1:50,000 scale initially, in a year's time. Topographical data or maps in digital and analogue form on WGS 84 (World Geodetic System 84) was made available for general public without any restriction. Further, SOI used the state-of-the-art technology in data acquisition, management and dissemination. High-resolution satellite imagery *viz.* IRS, SPOT, IKONOS data were being used for updating of topographical maps. All these data sets were useful in developing spatial data infrastructure that facilitated the availability and access to spatial data at all levels including government sectors, commercial sectors, academia and citizens in general.

Reference Ellipsoids

Ellipsoidal earth models were required for accurate range and bearing calculations over long distances. Loran-C, and GPS navigation receivers used ellipsoidal earth models to compute position and waypoint information. The models were defined as an ellipsoid with an equatorial radius having polar radius. The best of these models can represent the shape of the earth over the smoothed, averaged sea-surface to within about one-hundred meters. Reference ellipsoids have been defined by semi-major (equatorial radius) and semi-minor (polar radius) axes. Other reference ellipsoid parameters such as flattening, and eccentricity were computed from these two factors. Many reference ellipsoids were in use by different nations and agencies.

Geodetic Datums

Geodetic datums define the reference systems that describe the size and shape of the earth. Hundreds of different datums have been used to frame position descriptions since the first estimates of the earth's size were made by Aristotle. Datums have been developed from a spherical earth to ellipsoidal models derived from years of satellite measurements.

Modern geodetic datums range from flat-earth models used for plane surveying to complex models used for international applications which completely describe the size, shape, orientation, gravity field, and angular velocity of the earth. While cartography, surveying, navigation, and astronomy all make use of geodetic datums, the science of geodesy has been the central discipline for these initiatives. Referencing geodetic coordinates to the wrong datum can result in positional errors of hundreds of meters. Different nations and agencies use different datums as the basis for coordinate systems used to identify positions in geographic information systems, precise positioning systems, and navigation systems. The diversity of datums in use today and the technological advancements that have made possible global positioning measurements with sub-meter accuracies requires careful datum selection and careful conversion between coordinates between different datums. Further, two different datums are used for horizontal and vertical coordinates – the horizontal one is an ellipsoid while the vertical one is geoid.

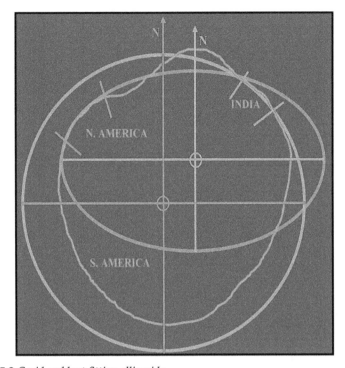

Figure 5.2 *Geoid and best fitting ellipsoids*

Geodetic datums can be broadly classified into two categories: (a) geocentric datums – World Geodetic System 1984 or WGS84, and (b) local geodetic datums – European datums 1987 or ED87 and Indian geodetic datum 1880. More than hundred local horizontal datums are used in different countries of the world, while GPS provided locations on WGS84 datum (Fig 5.2). With the introduction of GPS in the mapping and GIS activities, it had become imperative to establish relationship between the WGS84 and Everest datum (Fig 5.3). The *New Map Policy of 2005* had further indicated the necessity of such relation to bring out the Open Map Series or OSM. Further, the heights being provided by GPS refer to WGS84 datum. In order to obtain geoidal heights, the ellipsoid-geoid relationship is also to be established with a greater degree of accuracy (Chapter 4).

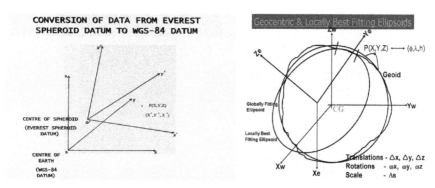

Figure 5.3 *A-Conversion from Everest to WGS84 datum; B-Geocentric and locally fitting ellipsoids*

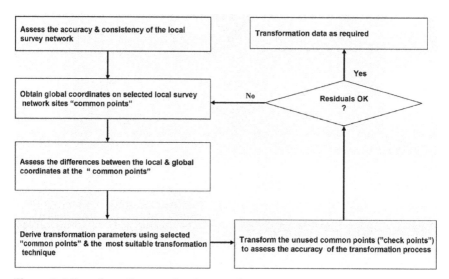

Figure 5. 4 *Transformation procedure between datums*

The transformation between any two horizontal datums required seven parameters: three due to shift of origin, three for rotation and a scale factor (Figs 5.4 & 5.5). These datum transformation parameters can be worked out using several points or stations in both the systems. Now software was available which can transform coordinates accurately between two geodetic datums.

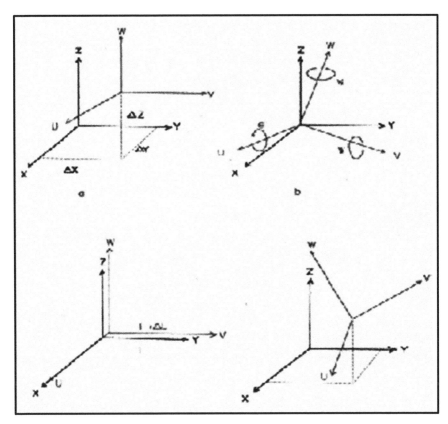

Figure 5.5 *Translation and rotations*

Indian Map Projection

Indian topographical maps have been based on polyconic projection. Due to colonial legacy, the neighbouring countries are also using the same projection. With the introduction of the New Map Policy in 2005, the *Universal Transverse Mercator (UTM)* projection was also adopted by SOI. The Open Map Series (OSM) is being developed on UTM projection. Hence, the polyconic projection and the UTM projections have been discussed here in some details.

Polyconic Projection

In the Survey of India, *Auxiliary Tables (1972)*, the details of the polyconic projection at 1:10,000, 1:25,000, 3:100,000, 1:50,000, 3:200,000 and *International Polyconic Projection* at 1:1,000,000 scales have been worked out. This publication also includes calculation for *Lambert Conical Orthomorphic*

Projection at various scales. The computations include length (in straight lines) for parallels, meridians, and diagonals for each graticule square for each map-scale. The calculations for polyconic projection have been the basis for the preparations of the Indian topographical map series.

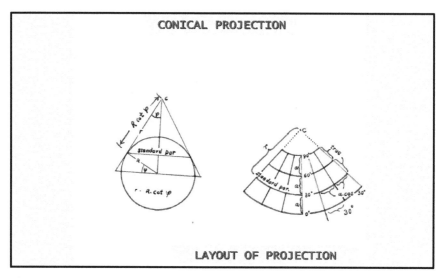

Figure 5.6 *Conical projection*

Polyconic projection belongs to the family of conical projections. It was probably developed by Hassler in 1820. It is called polyconic because several cones are involved in developing this projection (Fig 5.6). Here all parallels are without any distortion, *i.e.* the scale is correct along all the parallels. Further, scale is also correct along the central meridian. The parallels are arcs of circles but not concentric. Hence it is neither conformal nor equal area. Further, the central meridian and equator have been straight lines. It has rolling fit with adjacent topographical sheets in east-west direction. However, errors can be introduced if the adjoining sheets were to join. Obviously, there are problems in preparing seamless data in a rectangular coordinate system because each sheet is projected separately. Hence, one must take sufficient precautions if a large area is taken for GIS application.

Universal Transverse Mercator Projection

This projection has also been referred as *UTM grid system* as well. Based on the experience during the second world war, it was decided that all the North Atlantic Treaty Organization or NATO countries should bring their military mapping on to one system. For this purpose, *Transverse Mercator* projection was chosen as it was conformal projection suitable for topographical mapping. Further, in 1952, United Nations Cartographic Committee recommended that all topographical maps should adopt UTM grid system.

This grid system does not cover the whole world. It includes an area lying between latitudes 80ºS and 84ºN. The Arctic and Antarctic regions have been covered by the Universal Polar Stereographic or UPS projection (Fig 5.7).

In order to bring the distortions within acceptable limits, longitudinal portion are mapped on to the projection surface which are called as 'gores' having 6º longitude in width within 80ºS to 84ºN. Further, the 360º of longitude is divided into 60º of these gores, beginning at 180º west longitude and working eastwards these gores are numbered from 1 to 60º. They are known as *UTM grid zone* numbers.

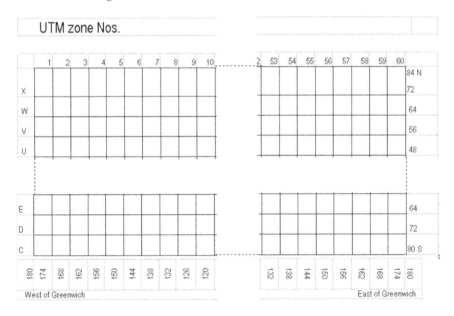

Figure 5.7 *UTM Sheet numbering system*

For locating any point on the earth surface, the UTM grid has been further divided in to 20 latitudinal belts starting from 'C' in the southern extreme to 'X' in the northern extreme. The letters 'I' and 'O' are omitted because they can be confused with 1 or 0. Each belt is having a width of 8 degrees latitude, except the northern most belt which is of 12 degrees latitude. The concerned area of the earth has been covered by 1,200 UTM grid zones, and each such zone are identified by the number of the longitudinal zone followed by the letter of the latitudinal belt, as mentioned above. In addition, there are following four elements in a UTM grid reference (Fig 5.7):

1. UTM grid zone

2. A pair of letters identifying a 1,00,000 metre grid square

3. An eastings value in metres

4. A northings value in metres

Since the central meridian is a vertical straight line and equator is a horizontal straight line, both crossing at right angles, it has been an ideal case for rectangular grids. The assumed northings (Y) are '0' metre is for the northern hemisphere and 10,000,000 metres for the southern hemisphere and the assumed easting (X) is 500,000 metres in both the cases. Since the maps are based on rectangular gird, it is easy for computer application and GIS as well (Fig 5.8).

UNIVERSAL TRANSVERSE MERCATOR PROJECTION

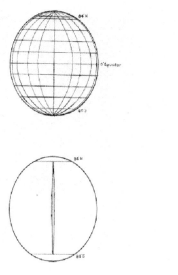
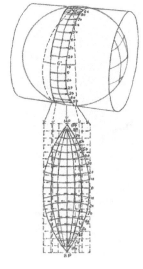

Figure 5.8 *UTM Projection*

Though the new series is on different datum and projection, it is possible to link sheet by sheet of the topographical maps of both the series (Fig 5.9). Hence with some calculations, the features of both the series can be compared. This was to facilitate in identifying geomorphological features over a period.

Chapter 6

Printing, Updation & Digitization

Printing of maps has been a regular activity of SOI particularly considering the requirements of the defence and security forces. In fact, such jobs use to supersede other priorities. Hence, there were huge setup for printing in the major centres of country, such as Dehra Dun, Delhi, Kolkata and Hyderabad. Such requisitions use to run into several thousand prints of a single map. As a result, a printing machine use to be engaged for weeks together to print one map. Due to this, the fresh maps or prints of the Department were kept pending for years in the presses. The originals use to get spoiled and heavy retouching was then required before they were again made print-ready. In several cases, the originals were sent back to the concerned directorates for this purpose. In the process the information use to get outdated. This was one of the reason for not providing maps on time. In any case, SOI continued provide the maps to defence establishments on priority basis.

Earlier, a government Lithographic Press was established in Calcutta (now Kolkata) in 1823. Nearly 56 years later, a section of this press was dedicated to the work of SOI for printing maps. However, this section was unable to meet the workload of the Survey. Hence, on 5th December 1851 it was decided to transfer the drawing branch of the Government Press to the Surveyor General's Office in Calcutta. In the following year, the Lithographic Press was installed. This in a way was the beginning of map printing in the SOI. Later printing of stamps were also carried out here. The first stamp was printed off on 4th May 1854.

Even though digital techniques were introduced in the organisation, manual methods still continued. Till recently, the department adopted scribing technique to be used concurrently with the conventional fair drawing for compilation of field data for publication of maps. Both the methods were skilled man-power intensive and technologically proven. Complete documentation were available for norms and standards applicable for these procedures. However, a number of such technologies in the recent past had been developed; as a result, some of them became obsolete. Some of the prominent changes were as follows:

- Flatbed plotters using pens and scribing tools and photo heads: These machines were heavy mechanical devices driven by computer and were expensive. However, precision was good. Nevertheless, by the time the technology was absorbed, the industry decided to discontinue production and hence no maintenance support.

- Drum plotters using pens and thermal plotters were phased out by the inkjet plotting technology. Pen, thermal and inkjet plotters could not meet the basic map printing requirements.

- Laser plotters have not been found economically viable for mapping.

- The available Map Publishing System was not deployed at production level. It was due to heavy equipment, very high cost, output not very fast, required skills, and overall it was not economical. However, it was having very high precision.

Printing & Reproduction

Conventional map printing activity was quite cumbersome in which map manuscripts were colour separated and press plates were prepared through a chain of intricate reprographic processes. The final printing was carried out on rotary offset multicolour printing machines. Map printing activities started in SOI when printing machines at Calcutta became fully operational in 1852. By 1865 another printing press was established in Dehra dun. The requirement of installing more printing machine increased considerably. During 1940, a full fledge printing press was placed at Dehra Dun which started functioning by middle of 1943. Over the years the reproduction techniques had undergone revolutionary changes. With the availability of outputs from digital cartographic data bases (DCDB), it became essential for SOI to install compatible printing machines. *Map Publishing System* and latest printing machines have now been augmented in SOI which were compatible to take on digital data also with desired accuracy and quality.

The then method of going through various stages during the map reproduction in bringing out the final hard copy map required a thorough review. The process of reproduction was changing from the then method of plate making to use new types of plates which worked out cheaper and time saving. Proof examination in reproduction office was to be dispensed with. Units responsible for data processing were made to complete the process and then submit the material for final printing in the presses. This was to reduce the movement of material between printing and processing offices.

There used to be a very heavy backlog for printing of the departmental maps. For example, in 2002, in South-Eastern Circle (Bhubaneshwar), there was a backlog of printing of about 200 maps. Before final printing of maps, at least two proofs were brought out which added to the printing load. Other directorates have more-or-less similar problems. Due to such a volume of printing jobs, there was a continuous requirement of manpower even if SOI focussed its attention to the departmental work. According to one estimation, there would have been no surplus staff in near future till digital techniques were introduced. Nevertheless, following plan of action was taken in this regard to meet the challenge:

(a) Since the defence requisitions involve large volume printing, new printing press in Palam under Director Survey (Air) or DS(A) was earmarked for this purpose.

(b) Four/five colour troll method was introduced instead of conventional printing having 11/12 colours.

(c) System of supplying of digital hard copies of maps was introduced by which maps from digital data can be provided on case-to-case basis. Price for such prints was already fixed. This was to reduce the time required for printing of maps.

(d) Proposal for temporary closure of unproductive unit, *i.e.* Photo Zinco Office (PZO) Printing Group in Dehra Dun was submitted to DST.

(e) The Barco Map Printing System was geared up for producing four colour separated for printing from digital data,

(f) Directors were empowered to finalize the printing of maps. Earlier all the maps used to get dumped in Map Publication Office (MPO)

Dehra Dun or in other presses.

The Western Printing Group (erstwhile 105 DLI Printing Group) started its operation in early 70's under the Directorate of Survey (Air) or DS(A), New Delhi. It became an independent directorate on 26th December 2003. This printing group was modernized after the installation of five colour Heidelberg Printing Press which was then considered to the state of art technology. It had complementary infrastructure including camera, proving press, cutting machine and automated plate processor. As on today, majority of quality printing work of the SOI and DST are being carried out here.

Proposal for Closer of Photo Zinco Office (PZO) Printing Group

As mentioned above, in 2002, the printing facilities were in Dehra Dun, Kolkata, Hyderabad and recently in New Delhi (Palam). These facilities were being used to print departmental maps as well as to meet defence (Army & Air Force) priorities. The printing facilities in Dehra Dun were under the Map Printing Directorate consisting of the following printing groups :

(a) No. 101 (HLO) Printing Group in Hathibarkala Estate,

(b) No. 103 (PZO) Printing Group in 17 EC Road, Karanpur.

An example of printing load in PZO Printing Group for three months in 2002 has been shown in Table 6.1:

2002	New edition	Reprint	Total	Total print
March	3	3	6	22,256
April	5	2	7	44,185
May	2	3	6*	40,378

Table 6.1 *Example of printing load*
Including one Miscellaneous map

In addition to above, this group was also engaged in printing forms and other stationary. However, such printing jobs could have been outsourced as well. Further, in order to run this establishment, following strength was posted as on June 2002 (Table 6.2).

Persons posted as on June 2002	Numbers
Group A	2
Group B	6
Group C, Division I	25
Group C, Division II	105
Maintenance Staff	13
Miscellaneous staff	25
Group C, Ministerial	8
Group D	89
Total	273

Table 6.2 *Postings in PZO printing group*

The sanctioned grant for the year 2001-02 for this printing group was about Rs 3.17 crore which was almost spent. About 92 per cent was disbursed under 'salary' head. Rest amount was paid for overtime allowances, office expenditure and the like. The above expenditure was for printing of about 75 maps in a year, *i.e.* Rs 4,00,000 per map (Rs four lakh) which was considered on higher side. This figure does not include other expenditure on infrastructure like electricity, telephone, cost and maintenance of the premises. Obviously a huge expenditure and manpower was not considered desirable. Hence it was proposed that the 103 (PZO) Printing Group should be closed. The immediate benefits were projected as follows :

(a) There will be saving in terms of electric, telephone and other maintenance expenditure.

(b) The Group C technical staff can be re-deployed in other printing groups where the machines were under used due to shortage of retouching staff, such as in No. 105 (DLI) Printing Group, Palam, New Delhi and No. 101 (HLO) Printing Group, Hathibarkala Estate, Dehra Dun.

(c) Group D staff could be redistributed in the field directorate where the progress in fieldwork was in slow pace due to shortage of such staff.

(d) Printing of forms etc could be out-sourced thereby promoting commercialisation. It was also be cheaper to do so.

(e) The surplus officers can be posted in other printing presses for better
management as already there was a shortage at management level.

Heidelberg Printing Machine

Considering the increase of printing workload due to new projects and for
the dual map series, it became pertinent to have a fully automated printing
machine. Hence a proposal for purchase of two Heidelberg Printing
Machines was mooted at a cost of Rs. 14.85 crores. The first Heidelberg
Printing Press was installed in SOI Campus in Palam, New Delhi. It started
functioning from 1st April 1999 (Fig 6.1).

Figure 6.1 *Heidelberg Printing Press, Delhi*

The other details were as follows:

- Cost of one Machine (1998) DM   2545455.00 INR   5.60 crore

- (Condition: Tender should be for two machines, but delivery
spread in two years)

- Contract with the Heidelberg is in DM

- 90 per cent payment made in 2000-2001 (March 2001): DM 2290910

- 10 per cent in 2001-2002 (1 DM = 22.47 INR)

Further in 2003, to ascertain reasonableness of the price, short listed companies were contacted to indicate their latest price:

- Man Roland (5 colours)12,45,000 Euro

- Heidelberg – 13,01,470 Euro *i.e.* INR – 7,02,79,380/-

However, it was recommended to go for the fresh, open and global tender. Nevertheless, printing machines for the following reasons were considered essential for delivery of maps to all stakeholders including Ministry of Defense:

- To maintain reasonable printing capacity for any operation

- Machines available with SOI are very old and have become obsolete.

Estimation of the Quantum of Workload for Updating

In order to estimate the manpower required for printing, the workload in terms of topographical sheets at a scale of 1:50,000 scale became important. The workload also got doubled as there was a new initiative to bring out an additional map series following the New Map Policy 2005. Earlier survey carried out for this scale has been mentioned in Table 6.3.

Surveyed year	No. of sheets
1978-82	907
1983-87	513
1988-92	494
1993-97	154
1998-2002	123
After 2002	09
Total	5,104

Table 6.3 *Survey in 5 years periods*

For bringing out the new series of maps, known as *Open Map Series* or OSM, not only conversion from hard copy to digital form was required, but also transformation of geodetic datum from Everest to World Geodetic System-84 or WGS-84 was necessary. India has been covered by about 5,100 topographical maps on 1:50,000 scale, out of which, about 3,100 sheets cover plain and semi undulating area; and 2,000 undulating and hilly area. Man-power required for such workload was worked out for sheets at different standard scales.

A. **1:50,000 scale**

Number of topographical sheets – 5,100

Number of days required to convert data – 5 days.

Total man-days required – 5,100 x 5 =   25,500 man-days

Average number of man-days in a month = 22

Therefore, number of man month required = 25,500/22= 1,160 man-months

Persons required to complete work in 4 months' time = 1,160/4 = 290

B. **1:2,50,000 Scale**

Number of topographical sheets – 396

Number of days required to convert data – 5 days

Total man-days required – 396 x 5 = 1980 man-days

Average number of man-days in a month = 22

Therefore number of man month required = 1980/22 = 90 man months

Persons required to complete work in 4 months' time = 90/4 = 23

C. **1:1M Scale**

Number of days required to convert =1 day

The proposal was to update these sheets as well. Regarding plain and semi undulating areas, remote sensing data from the Indian Remote Sensing Satellite or IRS-1C or ResourceSat P-6 was considered. However, for hilly areas, aerial photography was regarded as a better solution. Two exercises

were carried out during May 2003: (a) 57 A/9 – plain area in Karnataka; and
(b) 53 J/3 – undulating and hilly areas. It was difficult to identify even linear
features in hilly areas, however to some extent in stereo satellite imagery
proved to be useful. Aerial photography has been more suitable for hilly
areas. Considering the findings from such exercise, a proposal for updating
of topographical sheets using IRS-IC satellite imagery was submitted for
consideration. Parallelly, a proposal for procurement of aerial photography
from then National remote Sensing Agency or NRSA (now NRSC) was also
mooted.

The report of ground verification using remote sensing data gave some
indications about the workload for updating topographical maps at
1:50,000 scale. Earlier, topographical sheet No. 57A/9, covering the area
of Sindhnur, Manvi, Lingsugur *taluks* of Raichur district of Karnataka was
updated using IRS-IC satellite imagery (2003) and was used as input for
ground verification as well. Streams as inserted through imagery as well as
changes in alignment at some places were incorporated through imagery
agreed with the ground details. Expansion of existing village blocks and
roads as updated from imagery were also in agreement on ground. In one
or two cases, the road updated were not existing on ground since they
are found to be field bunds. Further, canals and tanks updated from the
imagery were found to be accurate while undertaking ground verification.
It took around 15 days to carry out updating a topographical sheet at
1:50,000 scale. Following conclusions were drawn from this experiment:

(a) Accuracy of linear features updated from imagery was proved beyond doubt.

(b) Existing of linear features extracted from imagery and its classification was needed to be ascertained on the ground.

(c) Linear features extracted from the imagery but not found on ground
may be due to temporal changes.

(d) Expansion of existing village block (area features) can be clearly extracted from imagery.

(e) Cultivation limits extracted from the imagery found to be correct on
ground.

The results were encouraging. Since then, the remote sensing data became at least an import input for updating of topographical sheets. The initial hesitation in using such data appeared to be withering away.

Challenges in Updating Topographical Sheets

The first recorded aerial photography in India was in 1927 on a scale of 4":1M. Since then the country were photographed several times and on various scales. These photographs have been considered to be most accurate source for mapping. Over the years photogrammetry and photo-interpretation have developed for a variety of purposes including geological studies and forestry. Then, apart from panchromatic photography, there was demand of infra-red, false colour and multi-spectral photography as well. SOI has been entrusted with the co-ordination and control of all work related to aerial photography. Some photographs have been cleared for educational purposes and a booklet entitled *Indenting Procedure for Aerial Photographs from Survey of India* has been issued (Fig 10.4B). However, the educational institutions were to take utmost care in handling these photos.

During the same time, experiments were also carried out for updation of such maps using aerial photography for hilly areas. However, additional objective was also to generate maps at 1:10,000 scale. Following five regions were considered for this purpose (Table 6.4).

Region	Area in sq km	Rate in Rs	Cost in Rs
Northern India	5,90,975	175	10,34,20,625
Northeast India	4,12,913	175	7,22,59,775
Gujarat	61,600	175	1,07,80,000
Southwest India	2,52,175	175	4,41,30,625
Near Jaipur	13,475	200	26,95,000

Table 6.4 *Updation of topographical maps with aerial photographs*

The total area cover was nearly 13.31 thousand sq km covering about 1,900 topographical sheets. The scale of photography was 1:40,000 or 1:50,000 depending on terrain height and maximum aircraft altitude possible. The standard photography specifications were taken: (a) overlap: 65 per cent forward and 30 per cent side lap; (b) B&W Kodak 2405 film or equivalent; (c) camera: RMK 15/23, (d) deliverables - B&W contact prints (2 sets), diapositives (1 set), and negatives (1 set). The average rate of photography was ranging from Rs 175 to Rs 200 per sq km. The total cost was estimated to be about Rs 23.33 crores. Then one photogrammetry scanner (very high resolution) was available in SOI, and one was proposed to be procured (approx. Rs. 70 lakhs). On the other hand, the IRS 1C satellite data per sq km was much cheaper even in comparison to Ikonos or Quick Bird data.

In order to understand the magnitude of such a herculean task of updating topographical sheets, some examples have been cited here. The first edition of the oldest surveyed sheet (No. 78G/6) of 1860-61 was brought out again in 1945. On the other hand, the 3rd edition of the latest surveyed sheet (No. 58A/12) was produced in 2002. Fig 6.2 indicates the status of the topographical sheets at 1:50,000 scale around the year 2003.

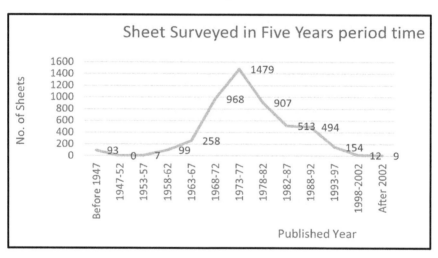

Figure 6.2 *Status of surveyed topographical sheets*

Traditionally, the updating of topographical sheets with the help of aerial photography had always been considered vital, hence an estimation was made for the cost considering different regions and a variety of projects that were in hand or likely to come to SOI for mapping. For updating of sheets at 1:50,000 scale and possible generation of maps at 1:10,000 scale for hilly areas, estimations were also made (Table 6.4). However, if we compare with the current prices of 2023, there are wide variations in cost, resolution, and other options.

Digitization

The progress in acquiring hardware and software for digitization from 1997 onwards has been shown in Table 6.5. From 2000-01, their number increased substantially but not good enough to make an overall transformation from analogue to digital. The number of such systems nevertheless multiplied with the establishment of the GDCs in different states and in other data centres.

Year	No. of systems	No. of software
1997-98	36	26
1998-99	20	10
1999-2000	20	9
2000-01	71	38
2001-02	136	100

Table 6.5 *Hardware & software for digitization*

From Table 6.6 it is apparent that there in the beginning there has been some progress of digitization of topographical sheets. In eleven years (1991-2002), out of 5,104 sheets at 1:50,000 scale, only 370 were digitized. Then the rate was only 60 per year; and with this rate, number of years required to complete the whole job was about 80 years. Nevertheless, the MO GSGS (Geography Section General Staff) had completed digitization of 693 sheets.

Scale	No. of sheets	Sheets digitized: 1991-2002 (11 years)	Sheets digitized: 2002-03 (1 year)	Sheets digitized: up to 1-4-2003	Balance to be digitized (9 months)
1:2,50,000	394	369	2	371	
1:50,000	5,104	370	676	1,046+693=1,739	3,365
1:25,000	19,526	287	617	904	

Table 6.6 *Status of digitization*

In order to shift from the conventional way of preparing maps, a massive exercise of computerization or digitalization was take up. Several meetings were held in the Union Department of Science & Technology, Surveyor General's Office (SGO) and in different directorates spread all over the country. Surplus staff from other trades were trained in digital techniques. Since the target was given to digitize all the topographical sheets within

December 2003, trainees from M/S NIIT and M/S Aptech were also engaged. There were only 569 digital systems were available and trained persons in SOI were only 707. Hence about 300 NIIT students were taken. The situation of digitization in October 2003 is displayed in Table 6.7.

Scale	Total sheets digitized up to September, 03	Target	Achievement	Remarks
1:250,000	371 & 3*	-	-	
1:50,000	3,760 618@	457	486	@ Digitized by GSGS
1:25,000	1,050	-	1	
DPMS/ State Maps	49	-	5	
Guide Maps	24	-	-	
Tourist Map	6	-	-	

Table 6.7 *Status of digitization as of October 2002*

The digitization of 1: 50,000 topographical sheets of border areas were done by the MO GSGS Division of the Ministry of Defence. The progress of digitization of such maps up to August 2002 has been shown in Fig 6.3.

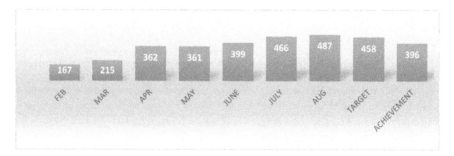

Figure 6.3 *Progress of digitization of 1:50,000 topographical sheets*

There were 5,104 topographical sheets at 1:50,000 scale. Out of which, up to September 2003, about 3,750 sheets were digitized. The remaining 1,344 sheets were to be digitized by the end of this year, and this was achieved. In order to demonstrate the digitization work and other GIS-related applications, the 70 Forest Party in the Hathibarkala Campus was attached with the Surveyor General's Office. The progress of this work during year 2003 in different circles have been shown in Table 6.1. Both the Digital Mapping

Centres located in Hyderabad and Dehra Dun made decent progress from January to August. Other circles which recorded good progress were Central, Western and Northern circles. The small set up known as SGO cell also digitized 142 sheets. MO GSGS of the Ministry of Defense also participated in this programme by digitizing 618 sheets. Further, in some circles, the achievements for the month of September 2003 were more than the target fixed. Obviously, they shared the workload of other circles (Table 6.8). The last sheet to be digitized was 91C/7.

Circle / Centre	Achievement During 2003								Total up to August	Targets for September	Achievement during September
	Jan.	Feb.	March	April	May	June	July	Aug			
NC	---	24	24	35	40	42	57	50	214	40	58
NWC	---	12	11	17	19	17	37	39	161	44	29
SC	5	14	4	12	18	20	28	24	144	25	19
SCC	---	10	10	12	21	25	26	21	128	25	33
SSEC	---	6	16	16	20	21	21	21	120	20	16
EC	---	---	10	21	31	39	41	60	198	40	08
NEC	---	---	8	21	24	31	35	28	146	30	23
SEC	---	5	7	18	24	26	35	44	144	40	45
CC	---	16	30	31	40	49	47	60	280	50	40
WC	---	26	27	42	38	42	53	51	244	50	53
DSA	1	3	5	9	5	4	19	29	72	20	20
DMC (D.)	2	11	21	25	19	18	20	21	277	25	22
DMC (H.)	13	29	26	36	30	36	20	17	353	20	13
MCC	---	---	---	1	11	13	9	11	123	19	11
SGO CELL	10	24	16	24	21	16	18	10	142	10	06
GSGS	---	---	---	---	---	---	---		618	---	---
TOTAL	34	167	215	362	361	399	466	486	3364	458	396

Table 6.8 *Progress of digitisation if different circles*

Note: WC: Western Circle; CC: Central Circle; SC: Southern Circle; EC: Eastern Circle; NEC: Northern Eastern Circle; SEC: South Eastern Circle; NWC: North Western Circle; MCC: Modern Cartographic Centre; NC: Northern Circle; DSA: Directorate of Survey (Air); GSGS: Geography Section General Staff; DMC (D): Digital Mapping Centre (Dehra Dun); DMC (H): Digital Mapping Centre (Hyderabad); SCC: South Central Circle; SSEC: South South Eastern Circle (PMP); SGO: Surveyor General's Office.

Digitization of topographical sheets also involved conversion from thick glass plates meant for different colours for transformation to raster and vector forms of topographical data. It was a herculean task as it required monitoring the process at several stages. Transfer of details from glass plates to films and then to generate separate digital files as required in GIS format were very challenging task. The glimpse of such process can be gathered from Table 6.9.

Area of responsibility	Total	Vector	Raster	Yet not started
CC	571	439	25	107
DSA	18	16	02	-
EC	320	290	01	29
NC	485	335	75	75
NEC	460	281	89	90
NWC	680	541	10	129
SC	557	296	144	117
SCC	421	316	42	63
SEC	460	225	118	117
SSEC	363	274	80	09
WC	829	588	88	153
TOTAL	5,164	3,601	674	889

Table 6.9 *Digitization as on October 2003*

The largest workload was Western Circle (829), followed by North-Western Circle (680) and Central Circle (571). Good progress was made by the different circles in vectorization of such sheets. However, in some circles, a good number of sheets were awaiting action in this regard. In total, for 3,601 sheets where data was generated, while for 674 sheets raster data was created. Only 889 sheets were yet to be taken up. The above numbers include sheets falling in (a) restricted, and (b) unrestricted areas of the country. In total, out of all the sheets, 2,551 were restricted, and 2,613 were unrestricted.

Circle	No. of system	Trained manpower	No. of NIIT students	Total	Monthly target
SC	35	43	--	43	25
SSEC	19	16	20	36	20
SCC	42	50	10	60	25
EC	16	26	--	26	40
NEC	39	45	--	45	30
SEC	47	90	--	90	40
WC	65	58	36	94	50
CC	54	87	49	136	50
NC	79	51	28	79	40
NWC	23	47	09	56	44
S (AIR)	26	18	--	18	20
STI	12	--	30	30	--
DMC (H.)	43	31	40	71	20
DMC (D.)	38	42	21	63	25
MCC	29	60	--	60	19
SGO's Cell	17	06	27	33	10
Total	584	670	270	940	458

Table 6.10 *Monthly target for digitization*

It is apparent from the Table 6.10 that the total monthly target for all the directorates was 458. In order to achieve this target, computers, technical staff and out-sourced staff from NIIT and Aptech were provided. Survey (Air) and the SGO Cell also shared the workload. Availability of limited number of trained man-power in some of the circles was affecting the progress as well.

Updating Topographical Maps

India, over the years, has generated a rich base of information through systematic data collection in the form of topographical surveys, geological surveys, soil surveys, cadastral surveys, various national resource innovative initiatives and application of remote sensing images. Further with the availability of digital topographic databases, high-resolution imageries and sophisticated techniques of data collection using GPS and tidal station, the accuracy of information content of these spatial data sets or map became very high.

The topographical maps on different scales were not meeting the requirement of planning and infrastructure development in rural and urban areas. Therefore, it was planned that SOI would provide digital spatial data on scale commensurate to the requirements of such planning. The methodology adopted in various stages of mapping was outdated. Latest methods of data capture and processing using digital techniques was the requirement of present day. In this context, all maps on scale 1:50,000 and 1:25,000 were available on paper form, which was cumbersome to use, need to be converted into digital form to provide a framework of spatial data to customers in NSDI platform. Topographical maps were needed to be updated using inputs like high resolution imaging data or rapid ground truth checking with digital techniques on a fast-track mode.

The data collected for ground truthing was to be done by field methods using the latest available techniques like GPS, total station, and digital level. In order to speed up the office data collection, the available inputs like aerial photo, satellite imagery, Airborne Laser Terrain Mapping (ALTM) data or other acceptable sources were to be used. Large scale aerial photographs on 1:5,000 scale was considered to be acceptable for mapping urban areas and 1:15, 000 scale for rural areas which was to facilitate generating digital data on larger scale with accuracy. ALTM technique was being tested to generate geospatial data for large-scale digital mapping. SOI was using mobile mapping techniques (palm top with GPS) for data collection instead of conventional plane table methods, so that the data updating and generating digital data can be done without much waste of time. Updating of existing 1:50,000 or 1:25,000 scale maps was considered as a huge task for the department. However, there has been a plan to update all these maps using high-resolution imagery like IKONOS, SPOT etc. SOI targeted to provide spatial digital data for existing 1:50,000 scale maps and make it available to stakeholders. It was also committed to provide precise control points at closer spacing using GPS duly connected with levelling datum and monumentation.

SOI also started to equip itself with GPS, digital levels and total stations immediately and start densifying the control. It was in the process of acquiring state of the art digital data capture and processing equipment and complete the generation of updated digital database on 1:50,000 scale by 2004. ALTM technology was proposed to acquire and acquisition of high-resolution digital data for urban areas. In the process, the department planned to switch over from producing large volume of hard copy maps on regular basis to providing such maps on demand or in soft copy form as per users' requirement.

Printing & Publications

Publication of topographical, geographical and other maps was carried out in the Map Publication Directorate which includes Map Record and Issue Office (MRIO), and was also responsible for certification of external boundaries and coast line of India. There were printing facilities in SOI premises in Delhi, Hyderabad and Kolkata. The published maps can broadly be classified as follows:

- Topographical maps

- Trekking Maps

- Special Maps

- Antique Map Series

- Discover India Series

- State Map Series

- District Planning Map Series

- Miscellaneous Maps and Publications

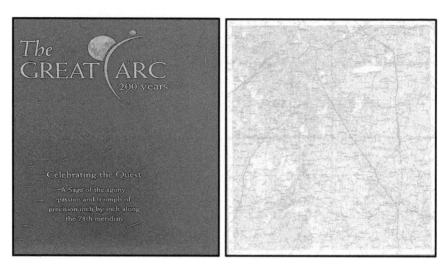

Figure 6.4 *Examples of printing items and publications*

Example of some of the printing items and publications have been included in Fig 6.4. Printing of both the series of topographical maps has been the basic job. However, there were regular publications being brought out from SOI as well. *Tide Table* is one of such publication. *School Atlas* has been a very popular item among the students (Fig 10.4). Its tourist and antique maps have been much in demand. DST also entrusts SOI whenever quality printing was required. Earlier, it had the reputation of printing stamps, look alike of the *Indian Constitution,* strip maps of border areas for the Ministry of External Affairs, maps and atlases of the National Atlas & Thematic Mapping Organisation (NATMO) and the like. Its press in Western Printing Group in Palam, New Delhi has the state of art facilities.

Chapter 7

Indian Spatial Data Infrastructure

During the autumn of 2000, a scientific delegation was sent by the Union Department of Science & Technology to Europe and North America to exchange experience and expertise regarding geospatial initiatives. This team consisted of Joint Secretary DST, Director NATMO and Director, MOGSGS of the Ministry of Defence. Executive Director of Technology Information, Forecasting and Assessment Council (TIFAC) of DST and the then Surveyor General of India could not participate in the delegation due to some pressing engagement in the country.

Geospatial information, including maps and images, were considered as a vital support to decision making at various level and implementation of action plans. With the availability of space borne imagery, global positioning system (GPS) data and geographic information system (GIS), users were able to process maps – both individually and along with tabular data and crunch them together to provide a new perception - the spatial visualization of information. Space based images have the characteristic to cover vast areas and have improved information quality as well as multi-spectral and repeat observation ability, thus, it was suitable for regional and global environment monitoring as well. The application of satellite images was also used for mapping and monitoring temporal changes in the environment.

The geospatial science encompassed a broad range of disciplines including surveying and mapping. The technology of digital image processing, the global positioning system and GIS were being used to integrate, processed and analyzed the spatial data for sustainable development of the natural resources. The digital data (*e.g.* high-resolution images, scanned aerial photographs, scanned maps and digital orthophotos) were then available for such purposes as well. GIS as a powerful decision-making tool became handy to cartographers as well. It began with a computerized topographic map as its base, GIS overlays and integration of graphic and textual information from separate databases. The end result was accepted as a customized and reliable tool to support decision making, problem solving and providing almost instantaneous answers to complex questions.

Particularly, from the discussions with the specialists in the United Kingdom, Canada and the United States, it became apparent that India needs to organize its geospatial data assets because in the digital and web world, there is no escape route. Hence, initially a National Task Force was appointed called as *National Geospatial Data Infrastructure* with the Surveyor General of India as the chairman. But these names itself lead to prolong discussions. *'Geospatial'* or *'Spatial'* was the bone of contention. Similar initiatives in other countries were consulted. Nevertheless, Department of Science & Technology and Survey of India were proposing for *'Geospatial'* while the Department of Space (DOS) and the Indian Space Research Organisation preferred *'Spatial'*. There were obvious reasons, based on the nature of data produced by these organisations, to plead for these terms. Finally, *'Spatial'* was agreed. Hence, the current nomenclature, *i.e. National Spatial Data Infrastructure* or NSDI, came into existence.

Early Considerations

Geospatial data Infrastructure was considered to play a crucial role in the development of the economy and promotion of industrialization at national and regional level. While being an advantageous for economy, infrastructural support was required for a highly complex and heterogeneous system. The country had varied expectations depending upon its socio-economic structure. Therefore, infrastructure cannot be merely considered as a physical foundation. These must be planned as a life support system for the man as well as his environment with milestones indicating infringement limits and related course corrections. The spatial data infrastructure (SDI) concept continues to evolve as it becomes a core infrastructure supporting economic development, environmental management and social stability in developed and developing countries alike (Williamson, Rajabifard & Feeney, 2003).

Following the recommendations of the visiting DST team, the *National Spatial Data Infrastructure* directorate was established initially in the SOI. During the deliberations of several conferences and workshops, it was

felt that this new arm of the Survey will be an interface with other institutions, other surveys and public at large. It was expected that then then
problems of the SOI will at least partially be circumvented by this new
initiative. Since, the DOS and its several affiliated offices were also dealing
with generation and application of geospatial data, they were also taken on
board. However, SOI and the DST continued their interest in establishing
this infrastructure.

NSDI has been a reality and not just a dream of few visionaries. NSDI was
aimed to play a key role in the developmental activities of the country
which had the requisite expertise, data, networking and other infrastructure. It offered tremendous opportunities as well. Geospatial scientists
and technologists were supposed to take advantage of the technology
and available geospatial information available through net, web and other
electronic media. The geographically referred information was in the form
of bits, bytes and pixels, and the volume of data was in megabytes, giga-
bytes and terabytes. This was to help in horizontal and vertical analysis of
smallest areas such a building in a town or a parcel of land in the village.
Further, the electronic flow of data was to influence geospatial information
in twenty first century by shifting (a) from so-called analogue theories to
empirical models; (b) enquiry from bottom up to bottom down; (c) developing indigenous models, and (d) revalidating established models, theo-
rems and postulates.

We live in information age which has become essential to solve problems
of national development. Further, spatial information was proved to be an
importance element underpinning decision-making for many disciplines
because it deals with location. In the sophisticated digital environment, it
has expanded to include databases, satellite positioning, communication
networks including internet, web technology and wireless applications.
It drew experiences from the geographic information systems, computer
science, land administration, geography, surveying and mapping, legal
and public administration disciplines. SDI initiatives in different countries
had a common objective, *i.e.* to create an environment or framework where
all stakeholders can co-operate and interact with each other to meet objectives at different political and administrative levels. SDIs were meant to
share data by avoiding duplications linked with generation, maintenance
and integration of data and development of innovative applications related

to business and national development. It appears that spatial information industry had evolved and will continue to grow. SDIs can become instrumental in having higher savings of all sorts.

Background

The development of NSDI should not be considered in isolation. It was a part of the major initiative taken up by the DST and the SOI. The other initiatives were the *National Map Policy,* dual map series, changes in field work methods, digitization of topographical maps and re-organisation of SOI. Though, the whole activity was to be a collaborative effort of all the stake holders, the question was who will host this initiative? DOS was claiming their stake in this regard due to its experience and contributions. The arguments not in favour of SOI were its style of functioning which was not always user-friendly. From both the sides, data standards and proto-type web portal were prepared. This led to a discussion which still continues in one way or other. Finally, based on the global experience, it was felt that the national mapping agencies should be made responsible for such initiatives. Further, by then, NSDI was included in DST in the *Allocation of Business.* Hence, SOI, an organisation of DST, was made responsible for this activity.

Nevertheless, additional issues were yet to be sorted out. There was no space available in Survey of India in Delhi which was preferred for such collaborative activity. Further, the office had to be centrally located and accessible to all the stake holders. Strict security measure was considered to have a dissuading effect. Fortunately, the Survey of India's Map Sales Office at Janpath was given a notice to vacate, and *in lieu* of that an office space in Ram Krishna Puram, New Delhi was provided by the Central Public Works Department (CPWD). This space did not fall under strict security cover and hence was considered to be suitable. Since then, the present premises hosts the NSDI office (Fig 7.1).

Figure 7.1 *Inauguration of NSDI Office in R.K. Puram, New Delhi. A Conference Hall named after Late Lt. Gen. K.L. Khosla, the first Indian Surveyor General of India is located here. (Col)*

Availability of space solved only a part of the problem. There were no funds and no *head* available in the SOI which was specifically to meet the expenses of NSDI. There was resistance within the SOI to provide funds as there were several other initiatives then undertaken which were depending on the limited *plan budget*. Somehow funds were managed from its budget for the initial years till the NSDI bill was approved by the Government. Some funds of DST through Natural Resource Data Management System (NRDMS) division were made available. On the other hand, office infrastructure was to be developed consisting of conference room, sever and other usual amenities. Ultimately, a conference room in this office was identified and was named after the first Indian Surveyor General of India. In addition to meetings, awareness campaigns among the stakeholders became necessary to circumvent the usual apprehensions. The data-producing organisations were requested to hold annual seminar of NSDI.

Accordingly, several such seminars were sponsored by the SOI, DOS, NIC, GSI, FICCI, IMD and the like. These events were held in New Delhi, Otty, Hyderabad, Agra and Lucknow. Such annual gatherings did help in building up understanding and confidence among the data producers.

NSDI Task Force tried very hard to involve the industry into their fold but the results were not commensurate with efforts. Representative of geospatial industry and industrial federation were always invited. However, industry had a different perception. Obviously, they took it as an opportunity to promote their business interest. Earlier their interest was in getting more projects and to sell hardware and software or to get their products and services approved. Almost all the standard software were then having facilities to exchange data from one format to another, or at least claiming to be so. Industry also looked towards NSDI for data quality certification which was not acceptable because it would have led towards bureaucracy and policing. Further, this was against the principle of NSDI, and DST and Survey of India preferred to stay away from it. Instead, NSDI encouraged the industry to set their own standards and promote the same while taking up any geospatial projects. Further, NSDI's effort to organize a forum of geospatial industry did not go very far. Nevertheless, recently, a successful attempt has been made outside the fold of NSDI.

Though the developments in the country regarding NSDI were happening with all constrains, there was a lot of encouragement from the international community. The British University of Durham, with their experience of handling European spatial data, extended full co-operation. Professor Michael J. Blakemore of this university visited India several times, particularly during the NSDI seminars. ITC, The Netherlands provided training to some of the Indian scientists in this regard. The experience of the United State's FGSI based in USGS was extended to India. Dr Julli Binder Maitra took special interest. The Canadian Natural Resources Centre not only helped India but also sent several delegations to India even headed by their Ministers. Several Indian delegations also visited Canada as well. Role of Professor J.R.F. Taylor has been commendable. He also headed the International Steering Committee on Geographic Data of which India is a participant. The then Surveyor General of India, was the chairman of the *Committee on Data Standards*. Another important international committee was UN/PCGIAP which was headed by Professor Peter Holland (Fig 7.2).

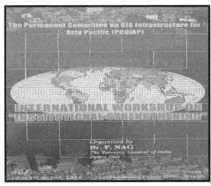

Figure 7.2 *PCGIAP Workshop in Mysore & GSDI Conference in Bangalore (Col)*

Here again, the Surveyor General of India chaired its *Committee on Institu-tional Strengthening*. Another Australian who helped India was Professor Ian Williamson of the University of Melbourne. Open Geospatial Consor-tium (OGC) and its representatives visited India to help in developing data standards and other facets of NSDI. The major event in this regard was the GSDI conference held in Bangalore in 2004. Overall there has been a demand for having a geospatial knowledge ecosystem (Fig 7. 3).

Geo spatial Activity	Survey/ Mapping/ Trained skill-workforce	Trained workforce for Survey/'Mappin, Geo-database and GISApps	Educated professionals for Survey/Mappin, Geodatabase and GIS Apps	Trained users development who would be users	School-level awareness
Present Availability Estimate	-15000-20.000	-eooo-ioooo	-800-1200	-25000-50000	NA
Estimated Additional need by end cf 2015	-20000 @4000- 5000 per year	-15000 @-2000- 3000 per year	-5000-S000@ -1000-1500 per year	-500,000 @ -50-100K per year	Estimated in phased manna-thru NCERT St3te School Boards
Knowledge/ Skill-inte rventions required	Industrial Training in specific Geo spatial Operations (2-4 weeks)	In-depth special-ised training in operations/ managing {3-12 months)	4-Year Graduate/ 2-Post-Grad-uate'PhD in Geospatial Technology thru University	UserTraining on specific GIS apps operations {1-2 orientation)	Basic chapters in 9-12 science curriculum; Additional GIS Kit knovsiedge exercises
Min Qual fcr knowledge' skill inter-ventions	10" OR 12" Grade sdiocl	Graduate in Science'Arts Or Diploma in Comp Apps	12" Grade leading to BTech" BTedi leading to MTedVPhD	Basically a Geospatial technology user in Centra I'State governments	Sdiocl at 6-SAND 9-12 Grade
Knowledge Credits	Professional Certificate	PG Diploma	BTech OR MTech OR PhD	Applied Certificates	Proficiency Certificate

Figure 7. 3 *Demand for a geospatial ecosystem*

Whatever the status of NDSI in India today is not very different from other countries. Though there is success in some sectors, much more is to be done. Meta data standards were found to be more acceptable and the web portals of the data producing institutions were getting ready. The conversion of spatial data to digital form followed the NSDI standards. Further, NSDI bill was approved by the Government and a separate division under the umbrella of DST was established. Perhaps more and persistent efforts and missionary zeal was required to promote co-operation in this high-level application of technology. As a result, NSDI came out of the folds of SOI.

The Indian Situation

Technological advancements in the fields of computers, satellite sensors, global positioning systems, geographical information systems and digital photogrammetry combined with the increased user aspirations in a way to obtain, analyse and apply the spatial data, had forced even the developing nations to consider mechanisms to make both spatial and non-spatial data accessible to users and to set standards. India has possibly the longest known tradition of systematically collected scientific spatial data, SOI have been doing so for the past 250 years or so. SOI is followed by India Meteorological Department (IMD) and Geological Survey of India (GSI) each over 150 years. Similarly, other departments such as Central Ground Water Board (CGWB), Forest Survey of India (FSI), National Bureau of Soil Survey & Land Use Planning (NBSSLUP) etc., have collected wealth of data over long periods of time. The NSDI partners have been with union and state governments and their agencies (Fig 7. 4).

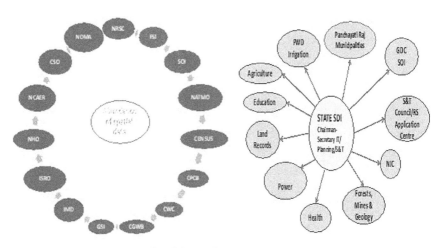

Figure 7.4 *NSDI partners and nodal agencies*

In the past three decades, DOS had produced a variety of data from various satellites. This data could also be utilized for developmental activities. Fortunately, in India, most of the spatial data is based on SOI topographic maps on scale 1:50,000 or 1:250,000. The analogue data when converted into digital format, it became available in machine readable formats. Despite the best conversion software available, there is loss of data when converted from one format to another. SOI has taken the lead in preparing a national standard exchange format for *Digital Vector Data* (DVD). However, it does not cater to all requirements. A lot of initiative was yet to be taken. Dr K. Kasturirangan has rightly stated:

> "I perceive that NSDI is a natural culmination of all that has happened in the area of surveying and mapping, remote sensing applications and GIS applications in this country over past years. NSDI is not a sudden "out of the blue" development, but a logical roadmap of what has been done and also envisions what needs to be done ahead. With a core information base, a spatial infrastructure is essential and this is where the NSDI becomes relevant"

(In Indian NSDI-A Passionate Saga, p. iv)

The institutions who have generated the data were highly possessive of the data generated and did not want to part with it. Even if they are willing to share, in the absence of metadata, it became difficult to know what is available.

Lastly, then policies did not permit digital data of topographical nature being utilized freely and there was a specific ban on putting the data on network.

Considering these issues, a Task Force was constituted by DST in November 2000 with Surveyor General of India as the Chairman. Members were drawn from various departments, organisations, users and NGOs providing spatial data. After resolving the initial hurdle, a National Task Force of NSDI was appointed again in 2002. However, this was not the end of the dilemmas. The Task Force met several times and had set up Working Groups for metadata, data exchange formats, network protocols, communication strategy etc. Three national and international workshops were held and the vision of NSDI was decided. Meanwhile, DST obtained all approvals for bringing out a second series of maps which could be made available without any restrictions. The task force had brought all the data providers to a common platform and they were expected to think alike. Thus, there was a better understanding of each other viewpoints and a consensus had emerged to evolve a National Spatial Data Infrastructure (NSDI). The data providers were engaged in the task of conversion of analogue data into digital domain and also to generate the metadata. The metadata servers were ready and launched.

Relevance of NSDI

India, had over the past years, produced a rich "base" of map information through systematic topographic surveys, geological surveys, soil surveys, cadastral surveys, various natural resources inventory programmes and the use of the remote sensing images. Further, with the availability of precision, high-resolution satellite images, data enabling the organisation of GIS, combined with the GPS, the accuracy and information content of these spatial datasets or maps became extremely high.

By then India had a well settled strategy, and the *NSDI Strategy and Action Plan* report was adopted in 2001 through the coordinated efforts of the NSDI Task Force. It was thought to be an important element for supporting economic and sustainable growth in the country. With the recognition that spatial information is a national resource, the citizens, society, private enter-prise and government had the right to access it. Only through common conventions and technical agreements, standards, metadata definitions, network and access protocols, the NSDI came into existence. While government was to provide the lead, the private enterprise, non-governmental organizations (NGOs) and academia had a major role to play in making it a reality. Further, it was desired to bring together the various "groups" and harmonizing their efforts in making this national endeavour a success. The design elements included standards, meta data, nodes, search and access protocols and clearing house (Fig 7.5).

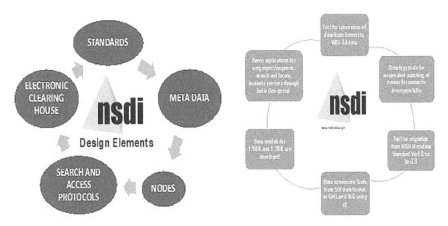

Figure 7.5 *NSDI Design elements, tools & technology*

Since the adoption of the NSDI Strategy in 2001, the NSDI Task Force had been working on developing a NSDI implementation mechanism by drafting an executive mechanism and preparing a possible NSDI bill. Then the NSDI vision was:

> "National Infrastructure for the availability of and access to organized spatial data, use of the infrastructure at community, local, state, regional and national levels for sustained economic growth."

It was decided that the NSDI be established under a legislation that would set out essential operational features including its composition and administrative structure, decision making powers, and the broad policy issues on which the NSDI would be entitled to make rules and regulations. The stated objective of the NSDI was the creation of a national infrastructure for the generation, availability of and access to organized spatial data to be used at community, local, state, regional and national levels to meet the requirements of spatial data for sustainable development. It was proposed that the NSDI be established as a statutory body, empowered to carry out various functions necessary to achieve the broad objectives stated above.

Major Constraints

Many planning departments and other national organizations such as Survey of India, National Remote Sensing Agency (now Centre) and National Atlas & Thematic Mapping Organisation (NATMO) have a long history of using topographical maps and remotely sensed imageries for preparing base maps for spatial planning. The methodologies of designs in such projects have invariably been worked out based on requirements projected by the resource users. The multidisciplinary approach to planning was often affected by non-availability of reliable, systematic and accurate data to all those involved in spatial planning. Several planning organizations had developed need based GIS to assist in analysing and making suitable projections. Non-availability of accurate and consistent spatial data base was the biggest constraint. There had been hesitation in standardization the database content on account of subjectivity in the requirement of data. Major constraints for standardizations of data were as follows:

(a) Geodetic framework and projections of topographical maps (base map) were often not in conformity with that of the cadastral maps.

(b) The information on the base map was generally outdated and does not contain the details required by the planners.

(c) The scale of map did not conform to the requirement of spatial planning at village or town level.

(d) Value added information on artifacts were to be collected from widely distributed sources and ground survey for items not depicted on map.

(e) Metadata *i.e.* definitions and description of parameters were not standardized.

(f) The mechanism for public participation depended on the local legal provisions and practices.

(g) Lack of technology upgradation.

(h) With no commitment available, the data producers were hesitant to invest in creating and maintaining the infrastructure for spatial database.

(i) Lack of legal commitment for dissemination of data to resource users.

The DST and the DOS came together and took an initiative to establish a platform for coordination and integration of activities between various data producing agencies to enable standardization, collection, compilation of data on a variety of related themes in a sustainable manner. Such a platform was also supposed to play a crucial role in formulating policies on dissemination of data and regulations for commercialization and privatization in various institutions of information technologies.

Major functions of NSDI

Initially, SOI sought solutions through for its problems such as better accessibility and providing maps and later digital data through this infrastructure. During its initial years, *i.e.* 2002-03, SOI adopted it as the main-stream activity. The speakers in the workshop held in Ramnagar projected similar views. After a series of discussions with the members of the Task Force, the major functions of the NSDI were identified as follows (Fig 7. 6).

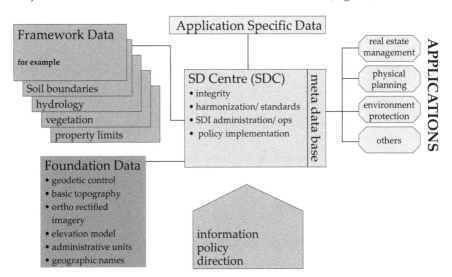

Figure 7.6 *Functions and applications (B&W)*

Major Functions

(a) To define policy concerning dissemination of spatial data to the government and private organisations.

(b) To identify and fix the standards of data to form fundamental data sets for topographical, cadastral and thematic applications.

(c) To lay down standards to which digital technology should conform and remain valid for a fixed period.

(d) Identify sponsors and formulate guidelines for funding for sustaining the infrastructure.

(e) Initiate standardisation of geo-referencing systems, plan for generating transformation parameters to support transportation of data in other popular geo-referencing systems and projections commensurate with the user requirements.

(f) To identify potential users, their resources and data requirements, economics of data transfer and legal cover etc., for setting up clearing houses.

Expected Results from NSDI were:

(a) Single window clearance for all users of geospatial digital data.

(b) Standardisation of geospatial digital databases including format, hardware, software, linkage, access and sharing and meta data.

(c) Finalisation of geospatial accuracy requirements for different users.

(d) More effective modelling for planning and development projects.

Issues related to NSDI

The issues related to NSDI were being formulated through various sub-groups which were deliberating with user agencies to assess their requirement of data set. Some of the issues were as follows:

(a) To lay down priorities and time frames for the development of infrastructure.

(b) To lay down standards for metadata in the fundamental data sets comprising of topographical and thematic data.

(c) To identify data generating organizations who can meet the standards of metadata and work on mechanism for transfer of standardized data to the fundamental data sets.

(d) To appoint sub committees to evaluate and scrutinize the on data transfer format, exchange format and other technical standards for the infrastructure.

(e) To identify the custodians of fundamental data sets, methodologies for its upkeep, updating, operational linkages and evaluation *vis-à-vis* meta data standards.

(f) Initiate standardization of geo-referencing systems and projection commensurate with the user requirement.

(g) To identify potential users, their resources and data requirements.

(h) To identify data requirements of users and lay down the contents of user data sets in consonance with the government policies on desensitization of spatial data.

(i) Policies on networking for access to data for browsing and value addition are defined.

There were many and varied stakeholders and they would include government at local, state, national levels (as both users and data collectors or owners), NGOs, educational and scientific organisations, private and public agencies. As of then, the major players were DOS, SOI, IMD, NATMO, GSI, NIC, Census Department and the like. Development of multi-user nodes was envisaged with NSDI portal (Fig 7.7).

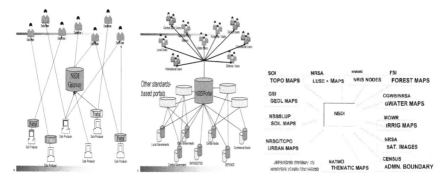

Figure 7.7 *NSDI data network and partners (B&W)*

Institutionalization of NSDI

The implementation of NSDI was major issue as it encompasses partnering ministries and organisations. In Ooty conference in 2002, a draft resolution was ready for discussion amongst the stakeholders. Further, after three years of efforts by the task force on NSDI, the following protocols emerged:

(a) A clear understanding on the next steps and the institutional framework – government resolution with an aim to have the act.

(b) NSDI Task Force to continue its effort in the direction resolved at Ooty

(c) NSDI agencies to commit specific programmes and resources for 10th Plan. Major thrust was to prepare metadata and convert analogue data into digital form.

(d) Each agency to design its own business plan.

(e) Private sector was to gear up for providing value added services and products to users.

Recognizing the need of accurate, consistent and updated geographical information, it was felt inescapable that the planners and decision makers should adopt an integrated approach for holistic planning aided with geo-informatics and arrive at rational solutions. Reliable and complete information on various themes was to facilitate better understanding of complex issues on conflicting objectives and to reduce the conflicts in decision making, especially where such divergence has its origin in disagreement on 'facts' and specially 'spatial awareness'. Such an initiative was supposed to go a long way in claiming g-readiness of to provide geographical information to resource users and industry for sustainable development of the nation.

Progress around 2003

After a lot of deliberations, some progress was noticed not only in DST or SOI, but also in the participating organsiation:

(a) Various central government departments like DST, DOS, Ministry of Environment & Forest (MoEnF), Department of Mines, Urban Development, Rural Development, NIC, Census Department, DAC, DARE, DOD etc. had initiated demonstration programmes towards NSDI activities.

(b) The finalization of the text of a NSDI Bill - which will be then to be processed by DST - first as an *Executive Order* and then the NSDI activities as an inter-agency programme.

(c) Drafting of following NSDI Standards documents - through inter-agency committees. The following documents were initiated and were in different stages of finalization:

(d) NSDI Content Standard - being coordinated through a committee and based on inputs from different agencies.

(e) NSDI Metadata Standard - being coordinated by a committee. Design and development of the Metadata Server also being taken up by DOS/NNRMS.

(f) NSDI Exchange (NSDE) Standard - being coordinated through a committee. The NRIS and SOI DVD formats are being merged into a NSDE Standard.

(g) NSDI Search and Access Protocol - being coordinated through the committee for NSDI Metadata.

(h) NSDI Networking Framework - being coordinated through a committee.

(i) NSDI Quality Standard – was then just initiated as a draft based on NRIS standards.

(j) NSDI Access Rules - being worked out by ISRO/DST.

The task force also developed a website and a NSDI Metadata Server which, when launched, was to provide information on spatial data availability to users. The *NSDI Metadata Server* is a catalogue of all spatial information of the NSDI agencies, as per a NSDI metadata standards.

Major Issues

The major issues which were very crucial to the implementation of NSDI were still to be considered. The utility of NSDI and its possible support to sustainable and economic growth in the country and for supporting developmental activities was well accepted by all. During 2nd NSDI Workshop held at Ooty in July 2002, the business potential of NSDI was discussed and it has emerged that the value of data and data products could be of the order of a few billion rupees. There was a need to evolve a business plan for NSDI as well as for the participating organisations. Further, the NSDI standard documents needed to be discussed, debated and, if endorsed, adopted for implementation. The details were available on the website.

Another major issue was the *Map Restriction Policy* – the review of which became essential for successful implementation. Actually the restriction on digital map data (topographical sheets and other maps; including datum and projection parameters restriction) were important issues for NSDI. The other issue was making the data available on networks for easy public access. In terms Ooty Communique, DST had prepared a note for discussion on policy issues:

> "There is a need for a NSDI policy that takes off from a modified and tuned map policy and defines the policy scope of each NSDI agency (SOI for topographical data, GSI for geological data, FSI for forest data and so on). The NSDI policy must address issues like - access rules, costing, copyright and licenses, value addition and the like. This issue needs to be debated in the NSDI workshops."

How should NSDI be implemented? As decided in Ooty, a draft government resolution was ready for discussion amongst ministries. Further, after three years of effort by the task force on NSDI, the following decisions emerged:

(a) A clear understanding on the next steps and the institutional frame-work – Government resolution with an aim to have the act.

(b) NSDI Task Force to continue its effort in the direction resolved at Ooty.

(c) NSDI Agency to commit specific agency programmes and resources for 10th Plan. Major thrust was to convert analogue data into digital and prepare metadata.

(d) Each agency was to design its own business plan.

(e) Private sector to gear up for providing value added services and products to users.

Role of Survey of India

In view of the present-day requirement, SOI planned to provide digital topographical databases for entire country on 1:50,000 scale initially within a year. Topographical data or maps in digital and analogue form on WGS 84 (World Geodetic System 84) was made available for general public without any restriction (existing topographical maps of SOI are on Everest datum). The department used the state-of-the-art technology in data acquisition, management and dissemination. Airborne laser terrain mapping (ALTM), GPS, electronic distance measuring (EDM) instruments, digital photogram-metry, modern printing technology and computer hardware and software were being used in map-making process. High-resolution satellite imagery *viz.* SPOT, IKONOS data were being used for updating of topographical maps. All these data sets were considered to be useful in developing spatial data infrastructure that was to facilitate the availability and access to spatial data for all levels including government sectors and commercial sectors, academia and citizens in general. Proposal to raise NSDI Directorate sent to DST and was immediately approved. As mentioned earlier, office order for raising NSDI directorate was issued in May 2002 by the Surveyor General of India which has been reproduced here (Fig 7.8).

THE GREAT ARC
200 YEARS
CELEBRATING THE QUEST

Hathibarkala Estate
Dehradun Uttaranchal
No........./NSDI/2002   **May 2002**

ORDER

RAISING OF NSDI DIRECTORATE

1. National Spatial Data Infrastructure (NSDI) stands created since 01
    Jan 2002 and its secretariat is functioning at Department of Science
    & Technology (DST) under the NSDI Division of DST. As direct-
    ed by the Hon'ble Minister for Science & Technology, HRD and
    Ocean Development, Dr Murli Manohar Joshi, the NSDI has to be
    made operational by 15 August 2002. A need has been felt to make
    the Secretariat fully functional by providing with the requisite
    statute, finances and manpower. All major data providers will be
    posting officers. As decided in the meeting of the Task Force held
    on 10 May 2002, a three-pronged strategy is being applied.

2. As a first part of the decided strategy, NSDI Directorate is being
    raised with immediate effect at New Delhi to service the NSDI
    secretariat and will be administered as given in succeeding para-
    graphs. This is being done with in the resources (Manpower and
    Budget) of Survey of India.

Location

3. The NSDI Dte will be located in New Delhi in the rented premises
    till such time NSDI acquires its own place of abode.

Organisation

4. This directorate will have all normal components of a specialised
    Survey of India directorates in skeletal form. The emphasis is to

have less manpower and outsource all services wherever possible. This directorate will have the following officers and staff posted from Survey of India.

(a) Director

(b) Staff Officer to Director – Gp 'B'

(c) E & AO – Gp'B'

(d) Assistants –2

(e) Steno/PA - 3

(f) Store/Record Keeper – Gp 'C'

(g) Technical Assistants – Gp'C' – 2 (Surveyor /P/Tr)

(h) MTD – 2

(i) Gp 'D' – 8

In addition to the officer from GSI who is already posted, officers will be posted from NHO, DOS, NIC, NATMO etc.,

Pay and allowances

5. Pay and allowances of officers and staff posted from Survey of India will be drawn by NSDI Directorate. However, the officers posted from other departments will draw their pay and allowances from their respective departmental offices located in New Delhi till NSDI gets its formal sanction of the Government.

Budget

6. Budgetary provisions are made out of the sanctioned grant of Survey of India under various heads and Director NSDI will be the controlling Officer of the budget. These allocations are being communicated separately.

Director

7. Head NSDI Division of DST will be the Director of NSDI director-ate as long he is an officer from Group 'A' cadre of Survey of India

Outsourcing

8. Director NSDI Dte will outsource the services such as transport, secretarial assistance, messenger, drivers, conservancy etc., wher-ever staff is not posted with in the budget allocated.

Hiring of accommodation

9. Hiring of office accommodation may be resorted to in the vicin-ity of DST as suitable space is not available in the DST premises. Proper procedure will be followed for hiring of accommodation.

Completion of raising

10. Raising will be completed by 31 May 2002 and established norms and provisions laid down in GFR, FR & SR will be scrupulously followed.

(Dr P Nag)
Surveyor General of India

Distribution:

1. Secretary DST
2. Head NSDI Division DST
3. Controller of Accounts, DST
4. CPAO, Dehradun
5. RPAO Jaipur
6. Internal: DSG I/II/III, ASG, DAF for necessary action and posting of personnel.

Figure 7.8 *Office order for raising NSDI directorate.*

User Applications and Solutions

While technologies such as GPS and GIS had matured during the 1990s, increasing attention was on the development of user applications and solutions. This technology segment was defined as software-cum-hardware solution to be developed specifically to solve problem of a geospatial information user. The critical challenge in this area was for the technology supplier to gain an in-depth understanding of the user's business environment so that the solution was optimized to address the user's key business issues. Built on top of the core technologies, user applications and solutions were developed using combinations of database management, object-oriented programming and systems integration tools and techniques. Clearly, geospatial data and GIS were considered to be extremely important technologies for addressing solutions in a number of applications. The process model of NSDI is shown in Fig 7.9.

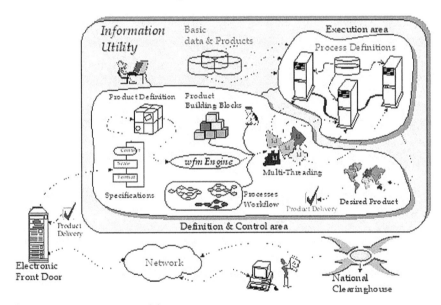

Figure 7.9 *NSDI process model*

India, over the years, had generated a rich base of information through systematic data collection in the form of topographical surveys, geological surveys, soil surveys, cadastral surveys, various national resource innovative programmes and the use of remote sensing images. With the growing awareness for information technology among users, an initiative was taken by DST in the form of NSDI. This soft infrastructure was the mechanism through which need for access to reliable, timely and spatial data were to be met. It was a first step in the direction of bringing together data providing agencies and to obtain commitment from them to make special data available and accessible to the user community, which conforms to accepted standards. The NSDI framework aimed at decentralized approach to achieve the following objectives:

(a) Develop and maintain standard of collection of spatial data.

(b) Develop common solution for discovery, access and use of spatial data in response to the needs of diverse and new user groups.

(c) Relationship among participant's organizations to support the continuing development of NSDI.

(d) Increase the awareness of the vision, concepts and benefits of the NSDI.

This framework when fully implemented was to be utilized for various developmental activities including preparedness, response and recovery purposes during disasters. The availability and use of spatial data was to have impacts on every aspect of society and will be made available to people who need them, when they need them and, in a format, so that they can make decision with minimal preprocessing.

NSDI at Global Level

At the global and international levels, SDI initiatives have been taken through different forums. The most important are being the Global Spatial Data Infrastructures (GSDI) conferences. UN sponsored Permanent Committee on GIS Infrastructures for Asia and the Pacific (PCGIAP), Permanent Committee for Spatial Data Infrastructure for Africa, Digital Earth conferences, United Nations Committee on Development Information (CODI), International Cartographic Association, International Steering Committee for Global Mapping, International Society of Photogrammetry and Remote Sensing and the like are contributing to the understanding of the SDI concept. At the national level, the Executive order from the President of the United States Office of 1994 was a great boost. On one hand, complimentary initiatives were followed in the US; while on the other, similar orders or initiatives were taken in other countries as well. The examples of the latter are Australia, Germany and Canada. With India's participation in global SDI-related activities, it became a part of hierarchy and SDI network ultimately converging under the aegis of the United Nations Global Geospatial Information Management (Fig 7.10).

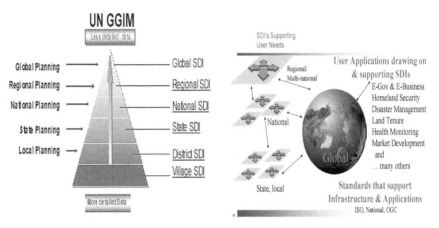

Figure 7.10 *Hierarchy of SDIs and network*

During February (2-6) 2004, the 7th International Conference on Global Spatial Data Infrastructure (GSDI-7) was held in Bangalore, India. The focal theme was *SDI for Sustainable Future*. This event was organised by the DST with SOI along with the DOS. It was held in conjunction with the 11th Meeting of the International Steering Committee of Global Mapping took place February 7, 2004. Further the Permanent Committee on GIS Infrastructure for Asia and the Pacific (PCGIAP) had its 10th Annual Meeting in Bangalore as well. For GSDI the meetings in Bangalore were the first administrative meetings after GSDI was formally established and registered in USA in 2003. GSDI was then considering moving its permanent secretariat from the United States to Bangalore, India. In tune with the global initiatives, a NSDI portal and SOI node for web services were later developed. This node provided data services over internet to desktop GIS users in open format, *i.e.* user defined format including OGC compliant and native formats (Fig 7.11).

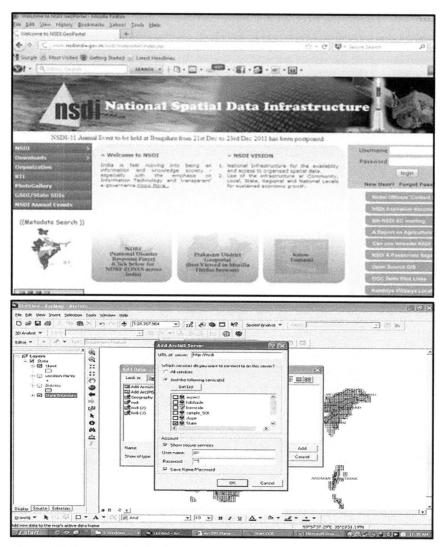

Figure 7.11 *NSDI portal and web service node*

Conclusion

The characteristics of the geospatial data set are changing. First and fore-most, in order to meet users demands effectively, the capacity for the real-time collection, synthesis and access must exist; data currency was thought to be essential. The data was to be scale less, seamless, without artificial boundaries, and linked to a time component that became critical to many applications, *e.g.* traffic flow management, routing and delivery, and tidal and marine traffic. Moreover, as technologies become more advanced, geospatial information was considered to be both more readily available and in greater demand.

There was also a growing trend toward the collection and integration of non-traditional data using secondary reference systems like voting, culture and housing patterns, gender, sales and industry. Furthermore, as technologies and applications become more globally used, geospatial data was to spread to and originate from non-traditional sources such as the voluntary sector, health councils, communities and peoples. However, regardless of what data are being collected by whom, unless they are easily and readily accessible, their value diminishes, hence there was a need for an exceptional geospatial data infrastructure. Furthermore, a well-developed national information infrastructure, enabling the dissemination and sharing of valuable, geographically referenced information, and with an ever-increasing audience of businesses, entrepreneurs, students and researchers, and communities, was widely accepted as an essential asset for any country to maintain and to advance its social and economic well-being. As such, geospatial data and the infrastructure in which they were organized can be considered to be a skill in its own right within the rubric of this technology.

Chapter 8

Research & Development Initiatives

It is interesting to explore the market potential of map products nationally and internationally, particularly keeping in view of the global scenario of land information and land markets. Further, in a changing world, the public sectors including national mapping agencies require business flexibility and the ability to react rapidly to change. Mentioned may be made to the technological advancements such as availability of high-resolution imageries, instrumentation such as GPS, total station and digital techniques which have opened up new opportunities of map-making and its updating process. Hence, in this context, the degree of readiness to transform geographical data being generated both in analogue and digital form become relevant.

In the past few years, SOI has been under criticism about policies for making their products available to the public. The restriction policies of Ministry of Defence of depriving public access to geographical data has also been a subject of discussion in many such forums. In order to overcome some the issues, SOI was able to come forward with alternate and path breaking solutions, such as National Spatial Data Infrastructure and the National Map Policy (2005).

Geographical data has been a fundamental requirement, which can influence and facilitate the planning and decision making at every level in day-to-day life. It is therefore relevant to analyse various aspects *viz.* scale, accuracy, standards and updatedness of such data being produced in analogue and digital forms. The need for suitable map coverage of a country has been well established for all organized activities of a modern state. For efficient administration, good neighbourly relations, internal and external security and for planned development of resources, need for timely availability of map on suitable scale and of desired accuracy were not only well recognized but well understood by planners and decision makers. Though the problem of providing such information in the form of maps or charts had been a stupendous and time consuming by itself, the subsequent problem of keeping this information up to date has been of equal magnitude and importance.

SOI has the responsibility of preparing topographical maps for India on various scales covering entire country and has been generating topographical data for over 250 years. These maps had all along formed the base for preparation of thematic maps and research. Besides topographical maps series, the department also brought out many maps to increase awareness of public towards the environment (Chapter 6).

Introduction of digital cartography and use of digital cartographic data base for bringing out updated maps was initiated in SOI in early eighties as an inhouse activity. Consequently, complementary developments associated with the use of digital cartographic database such as formulation of map data structure, development of data exchange formats translators etc. were also undertaken using the resources then available with the department. SOI takes pride in claiming that as a leader in this field, expertise of the department was extended to many scientific government departments and academic institutions and today these standards have been adopted by many departments. The digital data produced by SOI has been in use in many government departments and its undertakings after obtaining clearance from Ministry of Defence. While considering g-readiness of India with special reference to topographic data, it was logical that the issues related to dissemination of high-resolution geographical data in other fields such as hydrology, climatology, geology and seismology being generated by other government agencies were also considered for the purpose of maintaining consistency at the time of integration of data.

SOI provides national level maps and GIS to support various activities. Maps on various scales are being prepared in the department which show feature like natural and manmade features, and elevation in the form of contours etc. These maps could be of immense help in developmental activities including water resource management. Cartographic data in the form of maps, charts, atlases, digital data, GIS and other forms of geospatial data for designing and construction of various hydroelectric and irrigation schemes had been the prime requirements for the effective management of water resources. This resource is critical in Indian agriculture as well because more than 70 per cent of cultivated area needs rainfall. Excess of water or lack of water during critical cropping phase severely affects the yield. Therefore, management of water resources requires regular monitoring and inventory of surface and ground water potential.

Development in Surveying & Mapping

In the modern scientific term, *geomatic* is referred as an integrated approach of measurement, analyses, management, storage and display of the description and location of earth-based data, often termed *spatial data*. The data came from many sources including earth orbiting satellites, air and sea borne sensors and ground-based instruments. It is processed and manipulated with state-of-the-art information technology using computer software and hardware. It had applications in all disciplines which depend on spatial data, including environmental studies, planning, engineering, navigation, geology and geophysics, oceanography, land development and land ownership and forest. It had been fundamental to all the geoscience disciplines which use spatially related data.

Even two decades earlier, it was considered that we are in the midst of an information explosion, particularly our physical world and environment. Remote sensing even then could obtain spatial information in high-resolution imageries then better than 1 meter. Analysis and processing of large volume of information have been possible by the rapid advancement in the computer-aided technology. Users of topographical data as base map had realized the potential of digital topographical databases to improve the production of various themes and management of environmental data. The geoinformatics *i.e.* cartographic data was visualized as an information science performing a service function to a wide range of other disciplines like geology, meteorology, ecology, demography, geography, oceanography and the like.

The growing demand for economic well-being and for better quality of life had put stress on the management of resources. Therefore, the resource management was to be supported by an effective decision support system, which in turn required timely and high-quality spatial information. The science of surveying and mapping encompassed a broad range of disciplines including surveying and mapping, remote sensing, GIS and the GPS.

The latter was thought to be a powerful system for decision-making tool which was now in the hand of cartographers.   Beginning with a computerized topographic map as its base, GIS overlays and integrate graphic and textual information from separate database became possible.   The end result was a customized and reliable tool that can support decision making and problem solving and provided almost instantaneous answers to complex questions.

Himalayan Region

To study the seismicity and seismotectonic of Himalayan region seismotectonic cell was established in G&RB in December 1984. The activities of the project were to continuously monitor by the Project Advisory and Monitoring Committee (PAMC) constituted by DST in 1981. The cell was however formally closed in 1994. G&RB had also carried out repeat triangulation, levelling, gravity and geomagnetic observations in selected areas of Himalayan region. The studies were done in Kangra area and Garhwal area where long tunnels were passing through active thrust zones. The major active thrusts were Krol thrust, Srinagar thrust, main Himalayan thrust and Shali (Shanan) thrust. Application of geodesy for monitoring of crustal deformation in India dates back to 1920s when subsidence of various benchmarks was studied using very precise spirit levelling. Further, the geodetic studies for crustal deformation had been carried out in different parts of Himalaya. Few of them are discussed here.

A. Saharanpur – Dehra Dun – Mussoorie (Uttaranchal)

The Saharanpur – Dehra Dun – Mussoorie line had been levelled during 1861-62 and then repeated during 1903-04, 1905-07, 1926-28, 1974-77, 1991, 1992 and 1993-94 to analyse the stress pattern. From the analysis of these precision levelling data, the following conclusions were arrived at:

   (a) Subsidence just before, during and after (1 year) the Uttarkashi earthquake of 1991 indicates release of strain and one can say that the subsidence of this nature may be a precursor to future earthquakes.

   (b) Upheaval was due to unthrusting of Mussoorie block along Main

Boundary Thrust (MBT) with respect to Dehradun block which suggests that strain has again started accumulating rapidly in compressional regime between MBT and Main Central thrust (MCT).

B. Main Himalayan Thrust (Uttaranchal)

The crustal movement pillars were established across the Main Himalayan thrust near Uttarkashi in 1973. First set of triangulation and levelling observations were made during 1973-74 and repeated during 1977-78 and 1982-83. Fourth set of precision levelling was carried out during 1985-86. Repeat observations of 1977-78 have shown horizontal movements from 3 to 15 cm, and vertical subsidence was found to be 0.7 to 8.8 cm from 1973-74 data. Repeat levelling observations of 1985-86 had shown 1.4 to 20.0 cm subsidence over 1973-74 observations.

C. Srinagar Thrust (Uttarakhand)

The crustal movement pillars were established across Srinagar Thrust during 1973 in connection with Maneri Bhali Hydroelectric project stage II. The first geodetic triangulation and levelling observations were carried out during 1973-74. The scheme was repeated during 1976-78 and 1982-83. A fourth set of vertical data was collected during 1985-86. The observations of 1976-78 as compared to those of 1973-74 have shown horizontal movement of the order of 4 to 5 cm and vertical movement from 2 to 6 cm subsidence. For the next five years *i.e.* from 1976-78 to 1982-83, the horizontal movement ranged up to 14 cm in reverse direction.

D. Ganga Tear Fault (Uttarakhand)

Geological studies on the structure of Shivalik ranges had identified three main features *viz.* anticlinal axis, Bhimgoda thrust and Ganga tear fault near Haridwar. Crustal movement pillars were established across Ganga Tear Fault in 1974-75 and first geodetic triangulation and levelling observations were carried out during the same year. The triangulation observations were repeated during 1978-79 and the precision levelling in 1985. Further, horizontal movement of the order of 1.0 to 5.6 cm was noticed during 1978-79. Geodetic analysis shows that east bank of River Ganga had moved southwards, relative to the west bank. Strain pattern of the area indicated that the area got strained from one epoch to another.

E. Krol and Nahan Thrust (Himachal Pradesh)

Geodetic triangulation and precision levelling were carried out across the Krol and Nahan thrust in Yamuna Hydroelectric scheme Dakpathar, Uttarakhand. First set of observation was taken during 1973-74 and subsequent sets during 1975-76 and 1978-79. Strain pattern was also studied of this area based upon the repeat triangulation data. The results 1975-76 from 1973-74 showed the movements from 0.3 to 1.2 cm; and in the next three years, *i.e.,* between 1975-76 and 1978-79, the movements were of the order of 0.3 to 1.3 cm. Similarly, the vertical movement during the first interval, a subsidence of the order of 2 to 3.3 cm per year was detected while during the second interval an upliftment of the order of 0.9 to 2.9 cm was found.

F. Shali Thrust (Himachal Pradesh)

The crustal movement pillars were established across Shali thrust near Joginder Nagar during 1975. For recent crustal movement studies, geodetic triangulation and precision levelling were carried out during 1975-76 and repeated during 1977-78, 1979-80 and 1992-93. Movement in the southwest direction was noticed during 1977-78 when compared to pillar positions of 1975-76. During 1979-80 again the trend of movement continued in southwest direction with respect to 1977-78 positions. The result of 1992-93 when compared with 1979-80 also showed that the southwest trend continued except for pillar No. 6 which moved in northwest direction.

Jammu and Kashmir & Punjab

For monitoring vertical movement of the area repeated precision levelling was carried out on the following lines:

(a) **Pathankot – Dalhousie:** The levelling of 1960-61 was repeated in 1972-73 to find out the vertical movement of the area.

(b) **Pathankot – Dharmsala:** The levelling of 1909-10 had been repeated in 1918-19.

(c) **Jammu – Srinagar:** The levelling of Jammu and Srinagar line was first carried out in 1922 and was repeated during 1975-76.

From the observations, it can be inferred that there has been a general uplift of Himalayas in the above areas. Furthermore, the geodetic and geophysical studies for monitoring seismotectonic in Garhwal Himalayas were carried out after the Chamoli earthquake of March 29, 1999. Furthermore, in order to assess the post-earthquake effect and to monitor the seismotectonic activities in the region, following observations were conducted in 1999 (summer) and 1999-2000 (winter) field seasons:

(a) Gravity and geomagnetic observations in Garhwal Himalayas.

(b) High Precision Levelling on existing levelling lines.

    i. Saharanpur to Badrinath via Rishikesh.

    ii. Dehra Dun to Gangotri via Mussoorie.

(c) GPS observations in Garhwal Himalayas.

The above observations indicated that the stress developed within the crust due to 1991 earthquake had been completely released and the area in question had returned to normalcy by 1997. However, the repetition of above observations in same sequence should give some conclusive results.

Uttarakhand

Uttarakhand (then Uttaranchal) state constitutes the central Himalayan mountains and lies between latitudes 28.7° and 31.4° north, and longitude 77.7° and 81.1° east. It had dense forest area and was considered to be an abode of Gods and Goddesses and natural shelter for wild animals. It extends from River Tons, feeders of the River Yamuna in the west to River Kali in the east. Its northern limit is demarcated by Indo-Tibet water parity ridge and southern boundary districts are Dehradun, Haridwar, Pauri and Rudrapur. The Himalayan regions in the north constitute high mountains and glaciers. There are some well-known peaks *viz.* Nanda Devi, Trisule, Kedarnath, Badrinath, Chaukhamba, Bandarpunch and Kamet. This mountainous region is full of glaciers. There were more than 150 major glaciers located in the Uttarakhand region. Pindari, Gangotri, Milam and Khatling are some prominent glaciers of the region. It is because of these glaciers; the Uttarakhand region has been enriched with a network of rivers. The main River Bhagirathi originates from Gangotri glacier and Alaknanda joins at Devprayag to make it River Ganga. Yamuna originates from Yamnotri. River Kali is one of the largest rivers in Kumaun originating at Kalapani near Lipuiekh pass. Western Ramganga river originates from north slope of Dudhatoli ranges. With the network of major rivers in Uttarakhand, it has largest share of water resources than any other region of the country.

SOI beside preparing base map on various scale *viz* 1:250,000, 1:50,000 and 1:25,000 has prepared various utility maps for developmental activities in the region of the state. These are district planning maps, tourist maps, tracking maps and maps for various religious places. Some of other major activities of SOI for sustainable development of Uttarakhand were:

Developmental Project Surveys

SOI had accomplished preparation of large-scale maps for about fifty projects in Uttarakhand on various scales. Some of the major projects are: Joshimath Hydel scheme, Koteshwar Reservoir, Gauri-Ganga Dam site survey, Dhauliganga Dam survey, Pancheshwar Dam reservoir area, Pithoragarh magnesite project, Eastern Ramganga Dam scheme, Bhagirathi Valley survey, Nainital Town and surrounding, Srinagar Hydel scheme, New Tehri Township survey etc.

SOI had major role for implementation of national developmental schemes from planning stage to the completion of the project. It had established precise planimetric and height control in connection with planning and execution of various irrigation and hydroelectric projects. Geodetic and geophysical surveys have been carried out in stages to monitor the construction activities in recent past for the major projects located in the region. Maneri Bhali Hydel Project, Tehri Dam Project, Pala Maneri Hydel Project, Jamrani Hydel Project, Kalagarh Dam, Koteshwar Dam, Vishnu Prayag Hydel scheme, Dhauliganga Dam, Eastern Ramganga Dam, Srinagar Hydel Project etc. are some of the major project surveys taken up in the state.

Glacier studies

Glaciers play most significant role on the hydrology of Himalayan rivers. Scientific studies of glaciers assume foremost importance as they contribute to the development of water resources and understanding the ecological system. SOI had been contributing to the glacier studies for monitoring the extents variation, movement in horizontal and vertical directions and depth of glaciers. These studies contributed to the development of water resources, climate and weather predictions and understanding of the ecological system. The department had carried out studies in various major glaciers in the country including Dokriani Bamak and Gangotri glaciers during the recent past. These studies were taken-up as co-coordinated research programme on Himalayan glaciology through multi-disciplinary glacier expeditions. A special publication of Survey of India on *Inventory of Major Glaciers in Indian Himalayas* was also brought out during April 2000.

DST had evolved a national programme of GPS and other geodetic studies with special reference to the Himalayan belt for seismotectonic of the region. In this programme, there was a proposal to establish twelve permanent GPS stations in the northwest Himalaya. In first phase, two permanent stations were established at Dehradun and Pithoragarh. Continuous GPS measurements were to be taken up on all permanent stations. There were to be 200 semi-permanent GPS stations where 3-5 days observations were to be conducted at least once a year. Although GPS network of stations was mainly for geodynamic studies, but the user community in the region was to be benefited through this data for many applications related to developmental activities.

The Uttarakhand state came into existence on 9th November 2000; therefore, utmost attention was needed in the context of integrated development of this mountainous region. Special emphases were required for the development of agriculture, forest, tourism, hydropower and the like. The efforts in the form of base map data, seismotectonic and glacier study and GIS were to go a long way for the prosperity of the state. With the introduction of digital mapping, its topographical databases were of immense help in planning and development activities. The department was already geared up in providing digital data of the state for sustainable development in the field of watershed, forest and disaster management. Further, SOI had carried out various geodetic, geophysical and topographical analysis in the country including following glaciological studies:

- A team from Geodetic & Research Branch (G&RB) participated in multi-disciplinary glacier expedition to Chhota Shigri glacier from 1986 to 1988 for collection of geodetic data. It was sponsored by DST.

- G&RB also participated in Kolhai glacier (J&K) expedition during 1989, organized by National Institute of Hydrology. The aim of this expedition was to harness the water of River Lidder for irrigation and hydropower and also the estimation of spring season flow and utilization of Kolhai glacier to the flow of River Lidder.

- G&RB had again participated in multi-disciplinary expedition-1991, to Dokrani Bamak glacier in Garhwal Himalayas. It was also sponsored by DST.

- One team from G&RB was sent for Dokrani Bamak glacier to collect geodetic and gravity data in 1994.

- The Inventory of glaciers in Indian Himalayas was also prepared by SOI. In all 327 major valley glaciers for the present were identified – out of these glaciers, 60 are from Jammu & Kashmir and Ladakh Himalayas, 85 from Himachal Himalayas, 162 from Uttarakhand and 20 from Sikkim Himalaya.

Salient Features of Glaciological Studies

It was found that the average rates of movement in Chhota Shigri glacier were 3.7 m/year, 20.4 m/year, 32.6 m/year 42.8 m/year in snout, lower, middle and upper ablation zones respectively. Glacier movement in accumulation zone had been estimated 54 m/year. Movements in Dokriani Bamak glacier were of the order of 12.0 m/year, 27.6 m/year, 44.4 m/year and 27.6 m/year, in snout, lower, middle and upper ablation zones respectively.

Thickness of ice in Chhota Shigri glacier was found from 60 m near lower ablation zone to 104 m in accumulation zone. Thickness in Dokriani Bamak glacier was from 41 m near snout to 130m in upper ablation zone. The overall error to an uncertainty of ice thickness by gravimetric method was of the order of about ±10 metres.

Water Management

Management of water resources had been critical for the Indian agriculture because more than 70 per cent of cultivated area needs rainfall. Excess of water or lack of water during sowing period was critical for crop yield. Therefore, management of water resources required regular monitoring and inventory of surface and ground water potentials. Following cartographic products were considered essential for water resource management:

4) Preparation of flood area maps and river basins.

5) Ground water maps for drought area.

6) Mapping of coastal regions.

7) Irrigation system mapping.

8) Reservoir extent mapping.

9) Glacier mapping

Flood and drainage can best be managed by drawing block level plans through large-scale maps. SOI provided special maps on scale 1:15,000 in some selected areas. Indo-Ganga plains in Uttar Pradesh and Bihar and Brahmaputra basin in Assam valley had been mapped with contour interval (CI) ranging from 0.5 to 5m for flood control management. Surveys were also carried out on demand of Central Water Commission (CWC) and various other agencies, spread over several years between 1970 and 1990. These maps represented all types of features such as rivers, lakes, glaciers, canals etc. by specific symbols that helped in managing the water related parameters. Drought areas could be managed by well-planned strategies such as constructing the check dams and rainwater sheds with the help of large-scale maps.

In addition to graphic representation, detail reports were prepared by the field surveyors which includes sub soil information like water table, forest type, landslides, fauna and flora in the area. Although all these information were not depicted in the graphical form, but this information was available in the form of reports, which could be of immense use while generating GIS for managing and planning activities.

Glaciers have been playing a most significant role on the hydrology of the rivers. Scientific studies of glaciers assume foremost importance as they contribute to the development of water resource and understanding the ecological system. In the event of disturbances in the glacier regime, there can be droughts, floods and changes in ecological system. SOI had been contributing to the glacier study for monitoring the extent variation, movement in horizontal and vertical directions and depth of glaciers through geodetic and geophysical measurements.

In order to manage coastal region for sea water management and implementation of developmental scheme, SOI has carried out extensive work of coastal mapping extended up to 20 km from the coast from Nellore in Andhra Pradesh to Bangladesh border on East Coast on 1:25,000 scale. This task was undertaken on the request of the then Department of Ocean Development (DOD) in the context of study of storm, surges and cyclones. Sea level data generated through a network of tidal stations located along east and west coasts and in islands is being analyzed for the study of sea level variations.

Water constituted a major source of power generation without affecting the atmospheric parameters. This source of energy was in abundance in the Himalayan rivers. Engineering design of hydroelectric projects based on topographical information to channelize the river water through various civil structures like diversion tunnels, pressure shafts, penstocks and power houses and open channels were being monitored. SOI was playing an important role in execution of major hydroelectric power projects in India and abroad. The alignment of tunnels in desired direction was considered to be critical as the tunnels were to be excavated from two or more ends and finally, they required meeting each other. Further, the department successfully completed alignment and survey work including the mapping the extent of reservoirs for various major hydroelectric projects in the country and abroad. Survey for Chukha hydel and Tala hydel projects in Bhutan and Pancheshwar in Nepal were some of the examples of the geodetic survey by the department in the neighboring countries. In India, almost all major hydroelectric and irrigation projects were being taken up by the department for fixing the alignments of structures and mapping.

Crustal Movement Studies

The entire northern portion of Uttarakhand has been a high seismic zone. It has several faults and thrust lines which were responsible for landslides and earthquakes rocking the region frequently. Some of the major thrusts in the region include Trans Himadri Thrust, Main Central Thrust, Main Boundary Thrust and Shivalik faults. Main Himalayan Thrust, Srinagar, Krol and Nahan thrusts also exist in the region.

Geodetic studies were carried out using conventional terrestrial techniques by SOI which established a dense high precision geodetic control network of survey pillars, benchmarks and bases around the location of the active faults under investigation. Repeated observations over this network, carried out periodically; provided precise estimates of crustal deformation vectors and velocities, rotation etc. in horizontal and vertical directions between the observation epochs for various studies by scientific community.

SOI had carried out crustal movement studies in Main Himalayan Thrust, Srinagar Thrust in Uttarkashi area and Krol and Nahan Thrust near Haridwar. An earthquake struck Chamoli district in March 1999 followed by series of after socks. In order to assess the post-earthquake effect and to monitor the seismotectonic in the region, geodetic and geophysical observations immediately after the event were taken up. In this exercise gravity and geomagnetic observations were observed along the selected profiles. High precision levelling along profiles Sahranpur-Badrinath and Dehra Dun to Gangotri were carried out to assess the vertical deforma-tion. A network of GPS stations was also established to monitor horizontal movement in the area after earthquake (Chapter IV).

Interlinking of Rivers

Ministry of Water Resources had entrusted the work of survey of three link projects for preparing feasibility studies of links for transferring water from surplus basins to deficit basins. These were the time targeted projects, being executed on the directions of Hon'ble Supreme Court of India, the details of each project were as follows:

(a) *Sone dam – STG Link project*

Study area:Gharhwah and Palamau districts of Jharkhand and Aurangabad, Gaya, Nawada, Munger and Bhagalpur districts of Bihar.

Length: 339 km.

Requirement:Preparation of L- Section and cross section of 500 metres on either side of the canal axis, leveling height at 50 metre intervals by double tertiary levelling.

Contour interval:0.5 to 1 metre

Time schedule:Five months.

Estimated cost:Rs. 1.24 crore.

(b) *Farakka – Sundarban Link project.*

Study area: Mursidabad, Kolkata, Nadia & 24 Parganas of West Bengal.

Length: 64.3 km.

Time schedule: Five months.

Estimated cost: Rs. 76 lakh.

(c) *Rajasthan – Sabarmati Link project.*

Study area: Jaisalmer, Barmer, Jalor districts of Rajasthan and Banas-katha, Sabarkatha districts of Gujarat

Length: 525 km.

Time schedule: Five months.

Estimated cost: Rs. 256.65 lakh

Since SOI had to engage its available manpower for quick ground verification and updating of 1:50,000 scale maps, therefore it was proposed to
employ engineering students for six months during the field as proposed
by Punjab Technical University and similarly placed institutions at a
nominal stipend.

Airborne Laser Terrain Mapping

A pilot project using ALTM technology was taken up by SOI in collaboration with M/S Vandana Aviation Ltd, New Delhi. The objective was to
evaluate the suitability of the ALTM technology for generating high resolution Digital Elevation Model (DEM) and line maps on 1:2,500 scale with
0.5m contour interval. The selected areas were central Delhi and Dwarika
(Fig 8.1 & 8.2). The control points were provided by SOI and the flying took
place in May 2003. The data was processed in Survey of India premises in
Palam, New Delhi.

Figure 8.1 *Raw Laser data of Connaught Place, New Delhi*

Figure 8.2 *Ortho data of Dwarka area, New Delhi*

The deliverables were ortho photos and line map with contours and sport heights based on WGS-84 datum. Evaluation was carried out by a Board of Officers headed by the Additional Surveyor General, Survey (Air), New Delhi. The board made following recommendations:

- The technology was suitable for large scale mapping. However, it was recommended that initially projects on 1:5,000 scale with contours at 1 or 2m vertical interval (VI) be taken up till the operators achieve adequate experience and expertise, and only then large-scale projects on 1:1,000/1:2,500 scale with contours at 0.5 VI should be taken up.

- The sampling distances for the ALTM project should be commensurate with the accuracy requirement of the DEM and should preferably be less than the contour interval stipulated for the project.

   – SOI operators should be associated with the data processing specially at the stage of digitization details from orthophoto image as their experience will improve the quality of the line information generated from the ALTM data.

Naptha-Jhakri Hydel Project, Himanchal Pradesh

In the vicinity of Naptha-Jhakri hydel project in Himanchal Pradesh a lake developed due to blockage of the river water flow. In the event of this artificial lake gets busted due to any reason it would have affected the area downstream in this state and the impact would include areas near to the national capital Delhi (Fig 8.3).

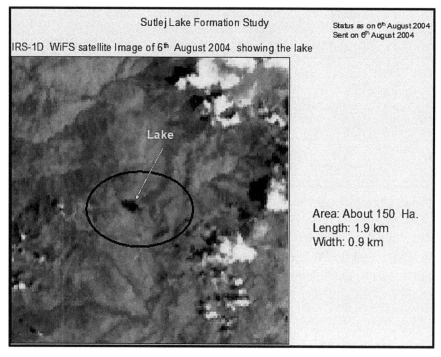

Figure 8.3 *Location of the blockage site*

Hence it was utmost necessary to monitor the conditions arising out of the blockage at River Pare-Chu. Further, the condition of the hydel project producing 1,500 MV power was also a matter of concern though it is located in the Jhakri district of the state on River Sutlej. The concerned area was nearly 3,880m above the mean sea level. This artificial lake was measuring approximately 2km X 1km (about 160ha). The upstream blockage was up to several kilometers. The concerned area falls in SOI topographical sheet No. 52L/11 (Fig 8.4).

Figure 8.4 *Lake formation*

Based on the topographical data, the water level in meter, area covered in ha and volume of water in cubic meters was closely monitored with the help of remote sensing data. The peak situation recorded was 3.9 thousand meters, 537 ha and 2.5 million cu meters respectively.

National Urban Information System

Ministry of Urban Development entrusted SOI to carry out detail mapping of 158 selected towns and cities in the country (Fig 8.5). The project was approved though it was a major deviation from preparing topographical maps at 1:2,50,000, 1:50,000 and at 1:25,000 scales. These maps were required by the urban planners, administrators, engineeers, utility providers, environmentalists and taxation departments. The objectives were as follows:

– To generate spatial data in terms of maps and image.

– To introduce use of modern data sources and methods.

– To develop and implement information systems concept to aid as a decision support system in planning and management of urban settlements.

Figure 8.5 *Location of towns and cities under NUIS project*

The method of creation of the urban information system was based on control points based on GPS and total stations, frameowrk on the then existing town maps, digital photogrammetric plotting or georeferencing of Ikonos or quickbird remote sensing data, field varification, attribute collection, GIS creation and finally printing of maps on various desired scales. The details are as follows (Table 8.1):

Definition	1:10,000 scale	1:2,000 scale
Projection	UTM	UTM
Datum	WGS84	WGS84
Heights	AMSL from 1:25,000 scale SOI maps in the form of 10 m contours	1 m contour interval
Spatial Contents (To be prepared using a SOI 1:25,000 scale base map.)	13 thematic layers (land use, geomorphology, slope etc.), 13 environmental layers	14 layers comprising of 80 features
Non – Spatial Contents	4 layers comprising of Demographic and Socio Economic data with village/ward as the spatial unit	Utility / Facility and Household attribute data
Target Application	Development plan, Master Plan	Municipal administration, management and utility GIS
Data source	Satellite data of 2.5 m or 1m resolution, 5 m multi-spectral	1:10,000 scale aerial photographs, stereo, B & W
Planimetric Accuracy	6 m accuracy as in 1:25,000 scale SOI base map	0.25 mm of scale for topographic maps = 0.6 m – 1 m
Frame Work	3' 45" x 3' 45"	1' x 1' (approx. 2000 m x 2000 m)

Table 8.1 *NUIS mapping details*

The parameters for the NUIS project were also based on the remote sensing (Table 8.2). Apart from the updating of some layers of topographical maps in GIS frame, this project was also used remote sensing data from various satellites and sensors in a big way.

Parameter (All values at 3σ)	1:10,000	1:2000
A] **IMAGE STANDARDS**		
Generic/Standard Resolution	▪5m XS or better	▪0.2 m
IRS Image Resolutions recommended	▪2.5 m Pan ▪2.5 m XS	▪Aerial Pan / XS
Projection for image outputs	LCC/TM	LCC/TM
Datum for image products	WGS 84	WGS 84
Image Frames (geometrically corrected; important for seamlessness)	3' 45"X 3' 45"	45" X 45"
Image Position (Planimetric) Accuracy (0.5 mm of scale) in m	5	0.5 (0.25mm of scale)
Band-to-Band Registration for XS data (0.25 pixel) in m	~1.5	~0.1

Table 8.2 *Image standards*

The mapping was to be carried out at 1:10,000, 1:2,500 and for utility (water or sewage) at 1:1,000 scales. The unit cost was then estimated to be Rs 5,000, 35,000 and 10,000 respectively. The total cost was projected to be about Rs 8,235 lakhs. Further, as a part of this project, several towns and cities were taken up. The example of the details in this urban information system for Delhi is shown in Fig 8.6.

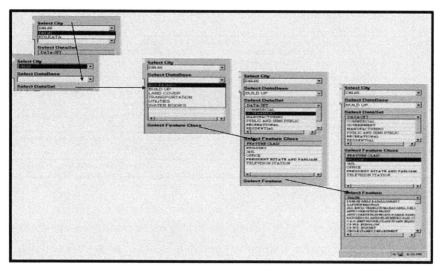

Figure 8.6 *Information system for Delhi*

The online customized urban information system is shown in Fig 8.7.

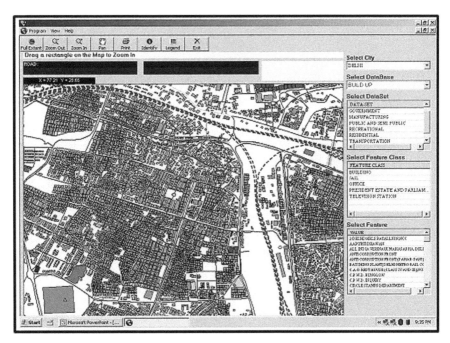

Figure 8.7 *On-line system*

The priority states were Gujarat, Haryana, Himanchal Pradesh, Jammu & Kashmir, Maharashtra, Madhya Pradesh, Odisha, Punjab, Rajasthan, Tamil Nadu and Uttar Pradesh.

Village Information System

SOI was equally responsible for providing geospatial information for rural areas. India has more than 6,00,000 villages located in different geographical regimes. These villages were lying in the then 600 districts which had been a common administrative unit throughout the country. The village boundary of each village was surveyed on ground and sometimes through photogrammetric instruments. Such boundaries were converted to digital form as polygons. Attributes were then attached with each village polygons as an important component of GIS or LIS. Query system was built as per requirement.

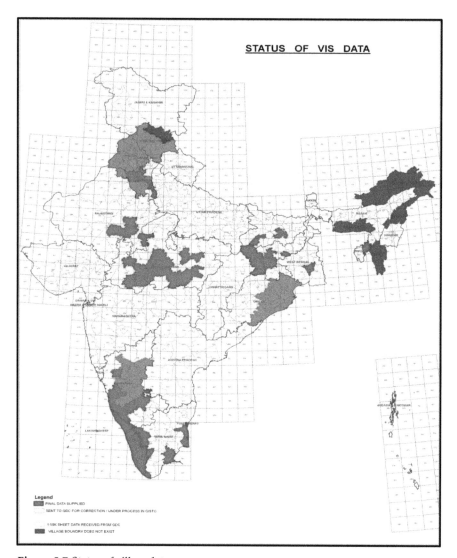

Figure 8.7 *Status of village data*

In case of selected district as per the priority (Fig 8.7), the village boundaries were verified in the field in Arc-GIS format with census attributes attached to each unit. The data of village boundaries as line were collected, sheet wise, from the respective state geospatial data centres (GDCs) in order to complete the whole district. After completing this exercise, the data was supplied to the National Informatic Centres by SOI. The example of Billary district of Karnataka is shown in Fig 8.8.

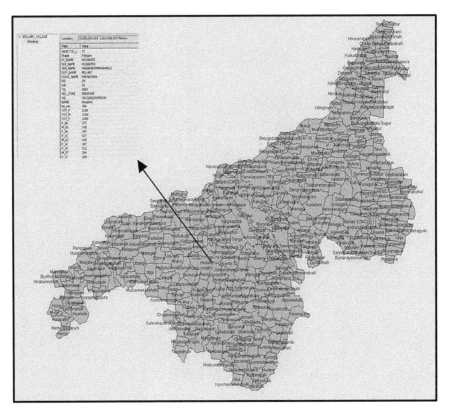

Figure 8.8 *Village information for Billary district, Karnataka*

Digital Elevation Model

There have been several applications either based on the existing topo-graphical data or based on exclusive surveys for specific purposes. For example, digital elevations models (DEM) were prepared for about 5,000 topographical sheets for which digital data was available. The contour and height data of these sheets were corrected and compiled followed by conversion of data to 3D and arc coverage which is then used for devel-oping DEM in IMG format with height attributes. The input data was contour and height in ".dgnFormat (WGS 84/UTM)." In short, the process was as follows (Fig 8.9):

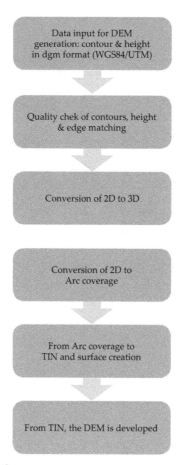

Figure 8.10 *DEM generation*

Survey of India Information through WEB

Several attempts were made to provide access information available with SOI to public. Hence, web sites were developed to provide information about the activities, products and services rendered by SOI. Search could be made by various ways:

(a) Topographical sheet number

(b) Latitude-longitude

(c) District name

(d) City or place name

Option for Hindi or English version was also made. Such site also met the requirements of NSDI which was then a major thrust areas in SOI (Fig 8.11).

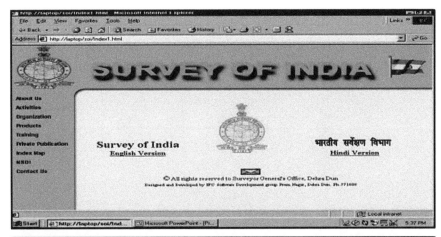

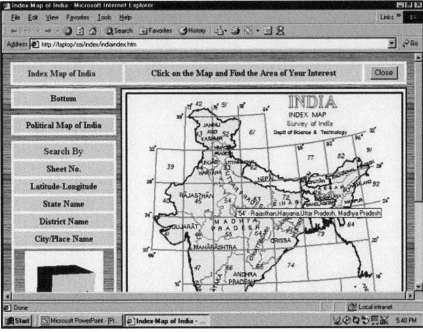

Figure 8.11 *Survey of India web site*

Conclusion

SOI had been actively involved in the geodetic and geophysical studies for the last two centuries and the accumulated voluminous data and analysed results were available in G&RB for the use of earth science community of this country. The collaborative efforts made under the aegis of DST for better understanding of earth, and its interior in the last few decades had paved way for clearer scientific understanding between various organizations. Setting up of National GPS Data Centre for archival, analysis and dissemination of data for seismotectonic studies at SOI had been one such venture in this direction.

Chapter 9

Commercial Arm

"Needless to say that executives of MNAs take a different
view. They claim that their organisations play an infrastructural role in the information requirements of government
and society as a whole and that the quality and reliability
requirements are such that their activities are a natural
monopoly providing a public good that cannot fruitfully be
carried out in the competitive marketplace…….In most cases
they will resist going further than contracting out of some
production work or some form of cost recovery in the sale of
their products and services."

Richard Groot (2000).

Department of Science & Technology was established in May 1971 with
an objective of promoting new areas of science and technology and to
play the role of a nodal department for organising, coordinating and
promoting science and technology activities in the country. An important
focus area of the DST was supposed to meet the country's requirements for
scientific data and information. Services provided in this regard include
surveying and maps for the ministries of defence, external affairs, state
governments, other developmental agencies and the private sector through
Survey of India (SOI) and National Atlas and Thematic Mapping Organisation (NATMO). It has been felt that for the economic development of
the country, scientific pursuits and commercialization have to go together.
Major scientific departments are now having commercial arms as well.
Hence, the commercial initiative cannot be done in isolation. DST had to be
taken on board along with SOI and NATMO.

Financial Conditions

It is important to understand the financial conditions of SOI during the
preceding years, *i.e.* prior to 2003. About 90 per cent of the annual expen-

diture use to spent on *Non-Plan sector* out of which maximum is on *Salary* head. The actual expenditure under *Plan* has been depicted in Fig 9.1. It has increased from below Rs 2 crores in 1991-92 to about Rs 5 crores in 2002-03 indicating substantial increase in the developmental works. However, this amount is not significant considering today's budget of 2023.

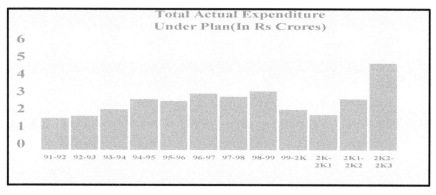

Figure 9.1 *Expenditure under Plan budget (B/W)*

On the other hand, the expenditure under *Non-Plan* steadily increased from 1991-92 to 1997-98. Thereafter, the expenditure remained almost constant (Fig 9.2). Though the salaries increased in all types of cadres alongwith pay commission reports, the reduction in number of officers and staff had compensated for it.

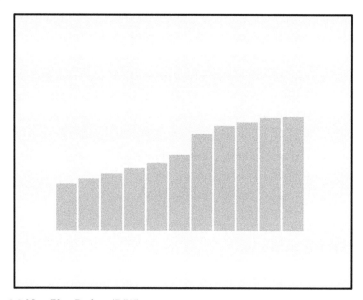

Figure 9.2 *Non-Plan Budget (B/W)*

The actual increase in salary which is mostly from *Non-Plan* budget shows similar trends (Fig 9.3).

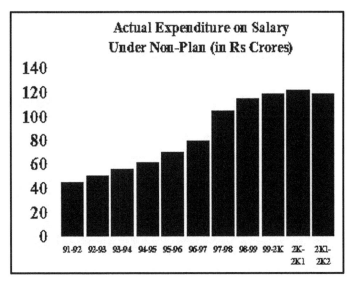

Figure 9.3 *Salaries under Non-Plan budget*

The revenue generated every year was about 20 per cent of the expenditure. During last one decade, the total revenue has rarely crossed Rs 15 crore figure, excepting 1999-2000 (Fig 9.4). A large part of the revenue came from the state government projects. However, the central government also made a reasonable contribution in revenue generation.

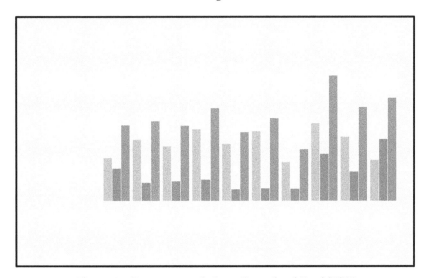

Figure 9.4 *Revenue earned: State, Central and Total (B/W)*

It may be gathered from the above description that the revenue earned from central and state governments projects was only a fraction of the expenditure. Some measures could have been taken to reduce the gap. The reasons were as follows:

- Uneconomic small units all over – consolidation required.

- Reasonable sum is paid for rents and taxes for hired buildings.

- Large estates are underutilized.

- Surplus staff in some sectors.

The then scenario could be transformed by introducing new technologies, new products, new partnership, new projects and new vision. Arrangements with the Indian industries could be a major step in this regard. An experiment was carried out with M/S Spatial Data Pvt. Ltd to bring out PC based interactive City Map Guide for Hyderabad and Bangalore (Fig 9.5).

Figure 9.5 *Example of joint venture*

It was a government-industry joint venture in digital cartography. Industry was encouraged to join hands with the SOI to harness potentials of both the systems. Unfortunately, the response was not always encouraging.

Possible Approach

Agencies like the SOI and the IMD had a long history of collecting data and developing basic products targeted at a limited set of users, primarily in the central and state government sectors. Today, new information markets have emerged comprising a wide variety of institutional and private users who would require basic and value-added scientific data and information generated by these agencies. However, because of the historic reasons, these agencies face several limitations in servicing these requirements:

(a) The data repository of these organizations, as well as the incremental data generated by them, was typically at a scale or resolution that was not suitable to meet the requirements of key user groups.

(b) The activity of product development to meet the needs of specific user groups required skills in value addition to base data of these agencies and would enable them to realize significantly higher revenues than were available for the sale of base data. Non-availability of these skills within the government setup had led to a large part of the new information markets to procure their data and information requirements from other data and service providers and hence deprived the SOI and IMD of significant commercial revenues.

(c) The SOI also lacked skills in marketing of data and information products to commercial markets, across areas such as institutional customer relationship management, branding and positioning of products for retail sales, distribution channels management, logistics management, etc.

(d) These agencies also faced significant resource constraints in addressing various commercial market opportunities in the form of technical manpower, technology for data collection and management, funding for large-scale data collection initiatives to overcome critical gaps in existing databases, etc.

With this backdrop, it was desirable for the DST to establish an organisational vehicle targeted at overcoming the limitations faced by its subordinate departments, the SOI, NATMO and the IMD, in servicing the commercial markets for various information products.

Situation in SOI

SOI had been engaged in the production and maintenance of various types of topographical, geographical and other public series maps covering India on various scales. Traditionally, the organization had seen its role as a geospatial data provider to meet the needs of the country's defence forces, and all its activities and products had been geared towards meeting this strategic requirement.

Then the volume of defence forces' requirement of SOI and products was diminishing, while the requirements of various new information markets were growing rapidly. Value-added geospatial data and services were required for diverse civilian activities including planning and development of infrastructure projects, disaster management activities, logistics management, delivery of economic and social services. However, the SOI's data base, organisational orientation, resources and skill sets available were not geared to service these information markets.

In the recent past, there was an increasing realisation within SOI of the need to refocus on various extra-defence market segments. A study was commissioned by the SOI in 2003 to examine the market potential and to highlight areas for change within SOI to address this potential. Key findings of the study have been highlighted in the following paragraphs.

Commercial Market Potential for SOI

The commercial market for SOI could be divided into two broad categories: (a) surveying and related services for the specific requirements of individual customers; and (b) off-the-shelf geospatial data products (such as urban maps), intended to be used by multiple customers. These markets were currently served by a relatively disorganised private sector at high costs and low quality. Nevertheless, the cumulative market size of the services business over a period of five years was estimated to be at Rs 1,200 crores. A large part of this market supported critical decision making across major infrastructure sectors. The significant market sub-segments included:

(a) *Roads* - A total of around 40,000 km of roads annually were under construction in India by way of new roads, lane extensions or strengthening the infrastructure. The market for surveying services for these projects has been valued at about Rs. 20 crores annually.

(b) *Power transmission* - The implementation of new transmission line projects called for preliminary surveys at the planning stage for route alignment through satellite imagery, followed by detailed ground surveys. It was projected that approximately 125,000 circuit-km of transmission networks was to be laid in the country in the 10th and 11th Plan periods. The surveying market potential from this industry was estimated at Rs 130 crores over a period of five years.

(c) *Power generation* - There was a significant requirement for quality survey in power projects. Space technology using satellite and aerial remote sensing data had a very important role in terrain mapping and scientific assessment of the ground conditions during preliminary investigations for site identification. Field surveys were also undertaken in the preliminary and detailed investigation stages. The market potential for surveying was approximately Rs 450 crores over a five-year period.

(d) *Urban Planning* – Large-scale maps were required for urban planning and management. The central Ministry of Urban Development, through the Town & Country Planning Organisation (TCPO), had initiated a scheme called the *Urban Mapping Scheme* to speed up preparation of maps on large scale for planning, implementation and monitoring. In the first phase, 136 towns and cities were to be mapped in the 10th Plan period; another 5,168 towns were to be covered during the 11th Plan period. This requirement translated into a potential market size of Rs 13 crore per annum over the next 3-4 years, increasing to Rs 120 crores annually in the 11th Plan period.

(e) *Water Supply & Sanitation* - Water supply and sanitation projects were an important market segment for surveying services. Significant survey work was also required across other application areas such as watershed development, rainwater harvesting, and hydrology modeling. These requirements amounted to a market potential of Rs 46 crores annually.

(f) *Railways* - Survey work for the Indian Railways typically pertained to route alignment, tunnel alignments and the like. Significant survey work was required in the construction of new lines (900 km per annum over the next 5 years) and gauge conversion (1,100 km per annum over the next 5 years). While the Indian Railways had a specialized survey department that undertakes surveying activity, around 30-35 per cent of their survey requirements were outsourced due to lack of internal resources. This has a market potential of Rs 3 crores per annum.

(g) *Industrial Estates* - Detailed land survey was required for infrastructure planning in industrial estates. Approximately 66,000 acres of industrial estates were planned for annually, translating into a market opportunity for surveying of Rs 6 crores annually.

(h) *Irrigation* - Survey work was primarily required for canal routing and contouring of nearly Rs 6,000 crores worth of irrigation projects annually. Around 50 per cent of irrigation surveys were being outsourced, the remaining being undertaken by state irrigation departments. The market potential for surveying from this sector was estimated at Rs 12 crore per annum. The cumulative revenue potential of markets for geospatial products then available in the market was estimated at Rs 700 crores over a period of five years. However, SOI had a negligible share of these revenues, despite providing a large proportion of the data that forms the base for these products.

(i) *City Maps* - Urban maps were important mass-market geospatial data products then available. The market for these products was estimated at nearly Rs 200 crores over a five-year period, and products sold by private agencies account for virtually all of the current sales.

(j) *Tourism Maps* - The tourism industry has been a large user of geospatial data, typically in the form of tourist maps and books. In other countries, these maps were typically prepared by government tourism or mapping agencies and play a key role in promotion of tourist flows. India, however, suffers from a critical shortage in quality product available in this area. The market potential of these products was estimated at over Rs 400 crores over the next five years.

(k) *Road network map*: Geospatial data relating to the country's transportation network, comprising of road network maps, was considered to be useful for varied applications such as route planning, logistics optimization and vehicle tracking and had the potential to add significant efficiencies to the country's road transportation sector. The market potential for these products was estimated at Rs 100 crores over the next five years.

Shortcomings in Addressing the Commercial Market

SOI suffers from several shortcomings that prevent it from addressing the available commercial market potential. These include:

(a) In the past the SOI was the sole organisation focused on collecting geo-spatial data for the nation. However, over the years the SOI activities have been shared by multiple organizations focused on specialised mapping requirements. As a result, the SOI's focus on and understanding of user needs had been progressively reduced, eroding its ability to meet user requirements over the years.

(b) The SOI had not kept up with changing market requirements for geospatial data. Data creation and updation had been slow and its core data base had been characterized by scale and high levels of obsolescence. Available data then covered the entire country on a 1:50,000 scale, and approximately 40 per cent of the country on a 1:25,000 scale. A large part of the requirements of the commercial market, however, are at scales larger than 1:5,000.

(c) Technology had revolutionized the entire value chain in the geospatial industry across the activities of data acquisition (use of GPS, palm tops, aerial surveys, etc.), data transformation (digitisation), data management and storage (data warehouses) and information dissemination (on-line, real time). The organisation continued to operate in a largely manual environment; efforts made to adopt and absorb technologies across the value chain were often constrained by limitations on budgetary funding support and restrictions on hiring.

(d) The Survey's total staffing in 2003 was nearly 10,400 people, down from 13,433 in 2000 following a recruitment freeze imposed by the government. There were severe staffing shortages in Groups A and B who have been responsible for core technical functions. Key non-technical functions such as marketing, information technology and product development also faced severe staffing and skill shortages.

(e) The SOI's status as a government department implies that it depends upon the Government of India's budgetary grants for all its financial requirements. The total expenditure in 2002-03 was Rs 153.79 crore against a budgetary allotment of Rs 172.33 crore. The budgetary allotment in 2003-04 was Rs 163.68 crore, a reduction of 5 per cent over the previous year. Funds available in real terms, after accounting for normal inflationary effects and the impact of pay commission recommendations, had reduced significantly over the last five years. The most significant impact of these constraints was on developmental activities across data updation, staffing enhancement and technology upgradation.

(f) Restrictions imposed upon the SOI under the *Restriction Policy* of the Government of India inhibit it from addressing several potentially large commercial market segments, even as other information suppliers in its industry often step in to fill the gap. Many of these restrictions have ceased to be relevant today in light of technological and market developments. These restrictions include:

– Survey maps at a scale larger than 1 million for certain areas of the country were classified as *restricted*. The areas falling within this restricted category included the entire coastal belt of the country up to 80 km inland, which housed a large proportion of all industrial investments in the country. This restriction significantly limited commercial activities of the Survey vis-à-vis the industrial sector.

– Restricted maps cannot be exported without the prior approval of

Ministry of Defence. Also, export of maps of unrestricted territories on a scale larger than 1:250,000, and microfilms obtained from such maps depicting any part of India including its international boundaries and showing topographical features by contours was prohibited.

 – SOI had large quantities of geospatial data available in digitized form, which was not made available to the private sector due to stringent restrictions on the use of digitized geospatial data.

 – Several restrictions were also applied to the dissemination of maps with coordinate references and height data.

(g) While many of the above restrictions inhibit SOI from addressing several potentially large commercial market segments, other information suppliers in its industry often stepped in to fill the gap. For example:

 – Topographical maps of India including the *'restricted'* zones, with coordinate references and contour intervals are available through Survey or government agencies in several other countries, such as the US Army Map Service Series U502, covering India and Pakistan. The 1:2,50,000 scale topographical sheets, which were restricted on account of security considerations along the external border and coastline, were available through agencies such as the Stanford International Map Centre, London. Satellite imagery, which provided various geographical and topographical information, are openly sold in bookshops in foreign countries while topographical sheets with similar information have been restricted in India.

 – Several private sector geospatial industry players frequently digitize SOI maps or maps of India obtained from other sources, value-add to these for a range of commercial applications and realise substantial revenues. SOI data, which often formed the backbone of these revenues, did not earn any amount by way of royalty.

 – Agencies such as the then National Remote Sensing Agency (now NRSC) deployed alternative technology for data generation and often capture significant market share across a range of commercial applications that fall within the SOI's domain.

(h) Government operating procedures constrained SOI's organizational optimization across operating dimensions such as procurement of capital equipment, hiring or staff redeployment, and incentive and even disincentive measures.

Commercial Market Targets for the Survey of India

The SOI could focus on addressing the commercial market opportunity initially focussing on delivery of high-quality services across areas, such as surveying with a progressive shift towards development and delivery of products as its base of geospatial data gets more aligned towards the needs of the commercial market segments. Further, the short-term commercial objective of the SOI would be to garner market share in surveying through provision of high-quality, cost effective and timely survey services, across sectors considering the following opportunities:

(a) Support the medium-term goal of product development such as urban planning and roads.

(b) Have the potential to generate significant service revenues such as power and water supply and sanitation.

(c) Sectors where the SOI would enjoy ease of entry such as railways and industrial estates.

Commercial sales targets for SOI from services were estimated to be Rs 25 Cr in FY 2005-06 increasing to Rs 185 Cr in FY 2009-10. In the medium term, the SOI was to focus efforts on product development based upon its existing data and incremental data generated through commercial surveying services. Key product segments have been mentioned above. Nevertheless, the long-term objective for SOI was to shift to products and the organisation would eventually seek to generate at least 70 per cent of its commercial turnover from product sales.

Interventions to address the Commercial Market

Several alternative organizational solutions had been considered for the SOI in servicing the distinct needs and requirements of the commercial market segment:

(A) Leaving the commercial market to private sector data providers (*i.e.,* maintaining *status quo*) was an undesirable option for several reasons:

    a.   Geospatial data needs of the commercial market were then being served by a relatively disorganized private sector at high costs and low levels of quality, driven largely by the SOI's lack of focus on servicing these needs. The SOI needed to step in to address this critical supply gap, especially given the mission-critical nature of various applications for such data.

    b.   Budgetary allocations for the SOI had been reducing, especially for developmental purposes. It had therefore become imperative to identify alternative revenue sources, so as to maintain the relevance and utility of its data across all market segments. The commercial market would not only provide the required financial resources for this, but it would also assist the SOI in addressing key data gaps across areas such as large-scale urban or suburban data.

    c.   Strategic needs for geo-spatial data had been on a continuous decline, evidenced by falling demand for SOI maps by defence agencies. SOI should have enhanced its focus on the commercial market to ensure its continued relevance as the national mapping agency.

(B) Addressing the commercial market through the existing SOI organization will require significant organisational interventions:

    a.   The SOI had been in its primary role as a service provider for the Ministry of Defense's geospatial data requirements, operated virtually without a marketing function. Organisational restructuring to include this function, and cultural changes to include a *'customer service'* mindset was felt necessary for the organisation which should be in a position to achieve its commercial market targets.

    b.   The organization had been highly centralized, and top management retains authority for most decisions. Directors at various directorates were not given sufficient administrative and financial powers as a result of which there was significant time lag in decision-making. The commercial market required decentralized structures that should have fewer management layers, reduced response times and greater employee involvement in operational decision-making.

c.   In the absence of commercial pressures in the area of service or product delivery, critical processes for coordination between marketing, sales and the SOI's field offices were under-developed. These required significant strengthening as part of the commercial market focus.

d.   The SOI has been traditionally collected and provided data to users, without any significant value addition. A commercial market focus required the creation of an organisational arm dedicated to product development and value addition to base data.

e.   The SOI had extremely limited focus on support functions such as information technology systems, human resources, finance and the like. As a result, these functions, which were critical to successful operations in the commercial market, required strengthening and development.

f.   There was inadequate staff at higher levels and more than adequate staff at lower levels in the SOI. A recruitment-freeze over the last several years had resulted in ageing of the workforce and the non-induction of several technical and managerial skills that were critical to meeting the needs of the commercial market.

g.   There were no incentives or disincentives to motivators for performance, which made it difficult to address without changes in the operating context of SOI as a government department.

(C) **Corporatisation of the existing SOI organization** to address the commercial market had also been considered. A corporate organizational framework should have assisted the SOI in addressing the needs of the commercial market in several ways – by imposing the discipline of meeting certain revenue and profitability targets; by lending the organisation operational flexibility; and by providing it with increased financial autonomy. However, corporatization of the SOI would have led to conflicts between the requirements of its role as a data provider for strategic and social needs, and the commercial imperatives that press upon a corporate entity; and hence it was then considered not desirable.

(D) Bifurcation (vertical separation) of the Survey into two organisations – a Social or strategic arm continuing to operate in a departmental format, and a commercial arm set up as a corporate entity was also considered. This option also suffers from some disadvantages that make it sub-optimal:

    a. Addressing the commercial market requires a mindset of customer orientation which the SOI then lacked. Mere corporatisation of the commercial operations is unlikely to infuse these values into the commercial organisation.

    b. Division of the SOI's field operations across two organisations will erode many of the existing strengths across areas such as technical capacity and lead to duplication of tasks and resources, eventually reducing organisational efficiency.

(E) Establishment of a full-fledged separate commercial vehicle for marketing activities was the final option considered. This vehicle would have focused on activities complementary to the operations of the SOI, including the development of high-volume geospatial products based on SOI data; marketing of geospatial products and services to commercial customers on behalf of the SOI; customer acquisition and relationship management; and the management of relationships with external agencies involved in activities such as software development and sales. SOI would have continued to perform all of its existing activities in servicing the strategic and social markets. It was also supposed to continue its core activities of collection, processing and management of field data.

(F) It was expected that entrusting commercial marketing and sales responsibilities to the commercial vehicle would have provided several benefits and enable the SOI to move towards a greater commercial market focus:

    a. Effective utilization of the scale and knowledge base of the SOI through commercialisation of available intellectual property.

    b. Improvements in the SOI's data base in terms of resolution of data and its currency.

    c. Improved ability to secure commercial business, given improved focus on commercial markets and greater flexibility in decision-making.

d. The possibility of incorporating incentive or disincentive mechanisms within the structuring framework of the commercial vehicle to ensure performance across all levels.

e. Separation of defence and commercial marketing organizations to eliminate any conflicts of interest and ensure the protection of the national strategic interest.

f. Greater flexibility in forging alliances, thereby improving abilities across areas such as service delivery, product development and distribution and sales of products

g. Financial benefits by reduced net outlays by the Government of India on operations of the SOI.

Public-Private Partnership

Public Private Partnership or PPP has become a slogan in every technology-based conferences or even in political discourses. It was considered to be the solution for all the problems including geospatial or GIS projects. The discussion under this umbrella often leads to arguments and blame-game. Hence, a serious effort was to be made to understand the issues involved in the so-called PPP model in GIS projects. What are the government sponsored initiatives and how the GIS industry has been appreciating or accepting the same? Will such efforts lead to better understanding or attract more projects or even more employment in the GIS industry? Such questions have become pertinent because GIS is not necessarily confined to the government institutions or the universities. Further, there is no specific discipline which totally patronizes this area of application. Hence, GIS is an open field which has several players, patrons and sponsors.

A better understanding could have been created by taking by certain measures, such as quick action by the public institutions for approvals and payments, continuity of policies and projects, and formalizing of exchanges of personnel as recommended by the fifth pay commission. This would have led to a better arrangement by appreciation of the strength and weakness of both the streams and how a new PPP model could have possibly been developed. Global recession may contribute towards this direction hence attention was to be given to domestic projects as well. Nevertheless, in the cost-cutting scenario in the west, perhaps the Indian GIS industry was still proven to be more competent. Finally, the co-operation between the public and private stake holders can potentially play a valuable role in bringing about sustainable development. India had proved its worth in the IT and ICT sectors, the GIS sector should not and must not lag behind. Manpower, technology, entrepreneurship and resources are available in the country to take leap forward.

Proposed Commercial Vehicle

The fulfillment of the commercial function for SOI would require a multi-disciplinary approach involving managerial capabilities across diverse areas such as business strategy and planning, marketing, sales and distribution, legal and contractual knowledge, financial structuring and management, production, information technology, alliances and partnerships to supplement each organisation's technical capabilities. The varied nature of the functional skills required would have made it difficult for the current set-up in the SOI to take on the additional requirements without seriously compromising the organisation's ability to perform its primary technical functions. Accordingly, it was suggested to set up a dedicated, whole-time group functioning as a distinct entity and having a legal and corporate identity of its own focussing on the activity of commercial dissemination of information products and services of DST.

Activities of the Commercial Arm

The *commercial arm* was to have following primary objectives:

(a) Promoting the use of information-led products and services across a wide range of commercial applications,

(b) Developing the Indian private sector's capabilities in the development, manufacture, dissemination and delivery of information-led products and services across multiple user segments, and

(c) Funneling of resources and market knowledge into agencies of the DST to guide and support technology research and data generation activities.

The commercial arm was to undertake the following primary activities towards the fulfillment of its objectives:

(a) Identification of specific initiatives and projects in the area of scientific information-led products and services

(b) Securing of data, rights for use of intellectual property and any other inputs required in the design, development, manufacture or production, marketing, sales and customer servicing in respect of the above products and services from the DST and its subordinate departments

(c) Product and service design and development, product manufacture, service delivery, marketing, promotion, sales, distribution and after-sales customer service, including outsourcing of any or all these activities.

(d) Licensing of product or service technologies for commercialisation by other government and private organisations

(e) Establishment of joint ventures (JVs) and other forms of association with other private and public sector agencies to undertake some or all the above activities.

The *commercial arm* was not normally to undertake product manufacturing, service delivery, or retail sales and distribution activities. These activities were to be promoted through mechanisms such as joint ventures, licensing and outsourcing to private organisations or to other operational arms of the DST. Further, given the wide range of its activities and high levels of technological flux, the *commercial arm* was meant to retain the services of technical and managerial experts on a regular basis with an objective of keeping its technical and managerial capabilities finely tuned to the needs of the marketplace and to constantly update its knowledge of technology developments.

To support its activities and to provide financial support to technology development initiatives of the DST, the *commercial arm* was to primarily generate resources from the commercial sale of various products and services, from the licensing and transfer of technology, and from dividends and capital gains in its joint venture investments. These resources would, if required by virtue of greater developmental activities than currently envisaged, be augmented through the route of market (debt) borrowings.

Promoters of the Commercial Arm

Given the technical nature of the proposed information products and services, and the need for strict quality control to fulfill requirements of the downstream mission-critical applications of these products, it was desirable for the DST and its subordinate offices to play an important role in guiding and controlling the activities of the *commercial arm*. At the same time, its operations in a competitive marketplace would require expertise across the wide range of functional areas discussed above that were not available with the DST or its subordinate organizations. Also, an important element of the *commercial arm's* operating philosophy was to include the generation of profits and operating surpluses, which had been an unfamiliar mindset to the DST and SOI. It was suggested that to inculcate this mindset and expertise at the outset, the *commercial arm* be established in joint venture mode with reputed external financial institutional investors, such as IDFC and IL&FS, who have been mandated to implement infrastructure projects on a commercial format which included solutions integrating geospatial data with information technology (IT) and information and communication technology (ICT). After having an in-depth study by Yola Georgiadou of ITC, The Netherlands and Girish Kumar of Survey of India in 2002, it was proposed to have a business model integrating GI and Geo-ICT (Fig 9.6).

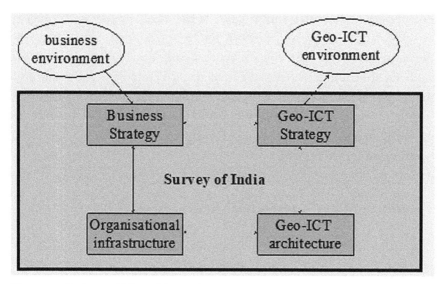

Figure 9.6 *Business of GI provision & Geo-ICT*

Organizational Form for the Commercial Arm

The *commercial arm* could take one of the following organisational forms:

(a) An autonomous body, such as a society registered under the Societies Act.

(b) A company registered under the Companies Act.

The society form of organisation had considerable advantages compared to a departmental set-up in matters such as autonomy in administration and personnel policies and procedures. However, several features of the society form of organisation make it unsuitable in the context of the *commercial arm*:

(a) There were inadequate regulatory controls on the operations of a society, across areas such as modification of its objects and the extremely limited liability of its members in their official capacity

(b) A society would not be a profit-making body and will therefore find it difficult to secure participation of the joint venture partners to raise resources from the market or financial institutions, or to promote downstream industrial ventures for specific projects.

A *company* registered under the Companies Act has similar advantages compared to a departmental set-up in terms of autonomy, operational flexibility and personnel policies. In addition, it would also enjoy advantages of access to external commercial resources and external risk capital to augment internal resources as required. Its operations would be subject to functional and regulatory discipline under the Companies Act, with requirements of annual presentation of audited accounts and the consequent accountability of its operations on a regular basis.

Given the nature of its activities, the *commercial arm* was to recruit high-quality manpower with extensive experience in the commercial marketplace. This would require the *commercial arm* to have the flexibility of paying market-driven remuneration and corresponding policies to enforce accountability for performance. It is therefore not desirable for the *commercial arm* to be subject to the guidelines of the Board of Public Enterprise (BPE). Accordingly, it was suggested that the *commercial arm* be incorporated as a Public Limited Company under the Companies Act, 1956, with the following shareholding:

(a) Government of India (trough DST): 26 per cent

(b) Financial institutional investors: 74 per cent

Linkages of the Commercial Arm with DST

The primary objectives of the *commercial arm* should be to catalyze the development and use of information-led products and services in various commercial applications. As a catalyst, the *commercial arm* will not normally involve itself or its personnel directly in activities such as manufacturing, service delivery, and retail distribution. In most cases, these activities would be performed on an outsourced basis or through joint ventures or through royalty arrangements, including such arrangements with the DST and its subordinate departments.

The commercial arm would therefore maintain close relationships with the DST and its subordinate departments in funneling commercial market opportunities to them and in developing and marketing information products based on the intellectual property of these agencies. In the process, the *commercial arm* would also pass on the following revenue streams to the DST:

(a) Charges for technical services procured by the *commercial arm* from the DST and its subordinate departments and offices on behalf of its commercial customers.

(b) Royalties on the use of the intellectual property of the DST and its subordinate departments in sales of various products developed and marketed by the *commercial arm*.

Organizational Structure of the Commercial Arm

The proposed *Commercial Arm* was proposed to be a public limited company. It was to have its registered office and head office in New Delhi. The board of the *Commercial Arm* was to have six directors apart from a chairman. The following composition of the Board was recommended:

(a) Secretary, Department of Science & Technology, Chairman

(b) Surveyor General of India

(c) Director General, India Meteorological Department

(d) Director, National Atlas & Thematic Mapping Organisation

(e) Representative, Infrastructure & Finance Corporation (IDFC)

(f) Representative, Infrastructure Leasing & Finance Services (IL&FS)

(g) Managing Director

The Managing Director was to be identified as a professional manager with adequate exposure to the commercialisation of information-based products and services. Further, the *commercial arm* was proposed to be a lean, professional organisation of technical, managerial and functional specialists, numbering about fifty people at initial stage. The key components of the arm's internal organisation were to be (a) research & development (R&D) arm, (b) operations arm, and (c) administration arm. The *R&D arm* was to focus on new product development. The *operations arm* was meant to be subdivided into *multiple strategic business units* (SBUs) focusing on activities such as marketing, sales, distribution management and customer relationship management for specific products or services. The *administrative arm* was also included to look after the usual managerial and organizational matters. The organizational structure of the *commercial arm* was also proposed (Fig 9.7).

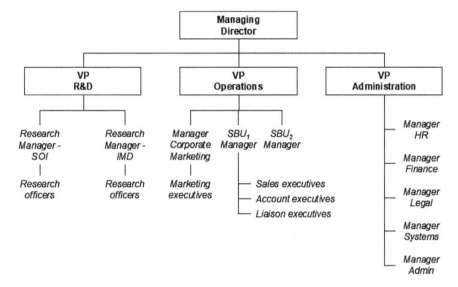

Figure 9.7 *Organisation structure of the commercial arm*

Financial Projections for the Commercial Arm

The commercial arm was estimated to generate revenues from two key activity streams:

(a) The sale of information *products,* developed on the basis of data resources of SOI, combined with data from other sources as required, packaged into standardized customer offerings for sale in large volumes. These products were to normally be sold through retail channels such as a distributor network or the internet. This arm was to pay royalty to the SOI for proprietary data of this agencies forming the basis of these products.

(b) The sale of *services* of the SOI, normally characterized by high transaction value, and customized collection, analysis and packaging of data to meet the specific requirements of the customer. These services should normally be sold through direct marketing to large industrial and institutional customers. Any contracts for service delivery arising out of such sales was normally to be outsourced to the SOI, with the commercial arm retaining a small margin on the contract value to cover its costs and overheads.

It was projected that the *commercial arm* will generate cumulative commercial revenues of Rs 940 crores over a five-year period up to 2009-10, of which it was estimated to pass on nearly Rs 670 crores to the SOI on an outsourced or royalty basis. Table 9.1 provides a breakup of the projected revenue streams.

	2005-06	2006-07	2007-08	2008-09	2009-10
Survey of India services	2800	4200	7800	12200	18500
Survey of India products	700	2400	3800	5900	8700
Total commercial revenues	**3,500**	**6,600**	**11,600**	**18,100**	**27,200**

Table 9.1 *Projected revenue (in Lakh Rs)*

Revenue Streams for the Commercial Arm

In the product development and marketing area, the *commercial arm* was to absorb all costs associated with value addition to the data resources of the SOI across areas such as product development, production, sales and distribution. It was also to absorb all costs associated with its direct marketing of SOI services. Based on these cost frameworks, it was anticipated that this arm will commence the generation of operational surpluses from the third year after its establishment.

(a) Projected Income Statement for the Commercial Arm

	2005-06	2006-07	2007-08	2008-09	2009-10
Revenues					
Revenues from product & service sales	5,040.00	9,180.00	16,260.00	25,380.00	38,190.00

	2005-06	2006-07	2007-08	2008-09	2009-10
Less pass-through to DST agencies	4,158.00	6,561.00	12,027.00	18,801.00	28,450.50
Net revenues	*882.00*	*2,619.00*	*4,233.00*	*6,579.00*	*9,739.50*
Other income	2.73	8.79	9.30	13.72	43.82
Total income	**884.73**	**2,627.79**	**4,242.30**	**6,592.72**	**9,783.32**
Costs					
Direct costs					
Production costs	420.00	1,440.00	2,280.00	3,540.00	5,220.00
Sales & marketing costs	352.00	609.00	963.00	1,469.00	2,259.50
Total direct costs	*772.00*	*2,049.00*	*3,243.00*	*5,009.00*	*7,479.50*
Overheads					
Product development expenses	300.00	250.00	300.00	500.00	600.00
Salaries	268.50	334.25	369.25	390.75	404.25
Administrative costs	219.81	252.76	272.17	285.48	295.46
Total - overheads	*788.31*	*837.01*	*941.42*	*1,176.23*	*1,299.71*
Total costs	**1,560.31**	**2,886.01**	**4,184.42**	**6,185.23**	**8,779.21**
Profit before interest, depreciation & tax	**-675.58**	**-258.22**	**57.88**	**407.49**	**1,004.10**
Interest	0.00	0.00	0.00	0.00	0.00
Depreciation	26.02	50.17	72.25	93.33	112.00
Profit before tax	**-701.60**	**-308.39**	**-14.37**	**314.16**	**892.10**
Tax	0.00	0.00	0.00	0.00	65.27
Profit after tax	**-701.60**	**-308.39**	**-14.37**	**314.16**	**826.84**

(b) Projected Balance Sheet for the Commercial Arm

Liabilities					
Equity	1,000	1,500	1,500	1,500	1,500
Reserves & surpluses	-701.60	-1,009.99	-1,024.36	-710.20	181.90
Debts	0.00	0.00	0.00	0.00	0.00
Total - liabilities & owners' equity	298.40	490.01	475.64	**789.80**	**1,681.90**
Assets					
Fixed assets - gross block	233.4	364.2	516	682.2	859.2
Less accumulated depreciation	26.02	76.19	148.44	241.78	353.78
Fixed assets - net block	207.38	288.01	367.56	**440.42**	**505.42**
Current assets - cash balances	91.02	202.00	108.09	349.38	1,111.22
Net current assets	91.02	202.00	108.09	**349.38**	**1,176.48**
Total - assets	298.40	490.01	475.64	**789.80**	**1,681.90**

Table 9.2 *Financial projections of the commercial arm (In Lakh Rs.)*

The above estimates include other data producing organizations under the umbrella of DST, including the then India Meteorological Department and the National Atlas & Thematic Mapping Organisation. Later, other divisions and institutions of DST could be included who are likely to extend products and services. The *commercial arm* would require upfront equity capitalisation from the Government of India and from financial institutional investors to support its establishment and operations for an initial period of two to three years. Financial projections of its operations, presented in Table 9.2, indicate requirement of an equity base of Rs 15 crores. Government of India was expected to contribute to Rs 3.9 crore as its share of this equity capital.

Chapter 10

Silent Presence

The silent existence of SOI is well known. It exists in the textbooks where there is international boundary or coastline. Its importance is felt wherever there are boundary pillars. Its premier product, topographical sheets, are seen in all the survey related institutions like Anthropological Survey of India or All India Soil and Land Use Survey or in remote sensing centres and even in state government agencies. Universities or colleges having geography or geology departments consider the topographical sheets as a prized possession. Same is applicable to aerial photographs which are also available from SOI. In the past, several attempts have been made to find out alternative to such sheets. Its presence is quiet but crucial. These maps do provide a snapshot of physical and cultural features at certain point of time with highest degree of accuracy.

Survey of India Offices

SOI offices are located almost in every state of the country. There are geospatial data centres, printing offices and specialized directorates all over. Some of the cities have several offices like Dehra Dun, Hyderabad and Kolkata. For example, Hyderabad city has GIS & Remote Sensing Directorate, Southern Printing Group, Survey Training Institute, and Andhra Pradesh GDC. The activities of such offices were secretive due to historical and security reasons. Supply of topographical sheets or tourist and city maps or antique maps or school atlas or even aerial photographs have been some of the items which schools and public in general would like to have from SOI offices.

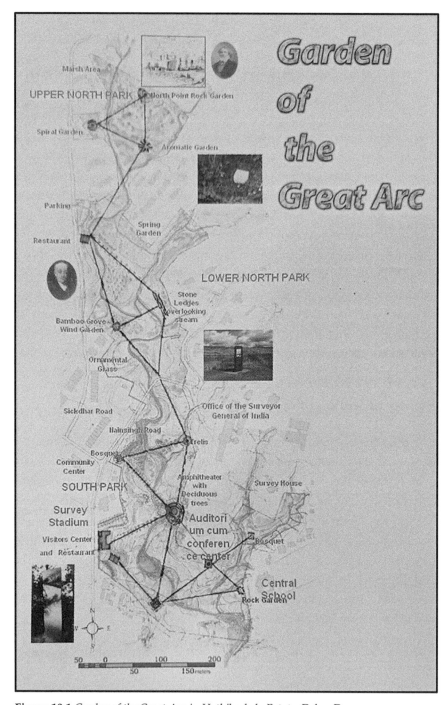

Figure 10.1 *Garden of the Great Arc in Hathibarkala Estate, Dehra Dun*

On the occasion of 200 years of the *Great Arc* series, SOI decided to develop its degraded land into a green corridor as an attempt towards the ecological regeneration of Doon valley in the form of the *Garden of the Great Arc* which was to commemorate one of the most significant scientific endeavours of the modern times (Fig 10.1). This garden is in Hathibarkala Estate, the headquarters of SOI in the city of Dehradun. Chief Minister of the then Uttranchal (now Uttrakhand) laid the foundation of this garden on 9th November 2002. It provides an aesthetic landscape for local people and also a recreation spot for the visitors. Further, the selection of species in the park was made keeping in view of several factors, *e.g.* aesthetic value, spiritual consideration, habitat requirement of birds, butterflies, animals and the like.

In addition, some of the survey offices have been in the centre of the cities. If permitted, common people would like to come for morning or evening walks. The Mussoorie campus at a hillock has the only playground of the city. The Meghalaya and Nagaland GDC in Bonnie Brae Estate is a well-known landmark in Shillong. West Bengal and Sikkim GDC is also a land-mark on the Park Street in Kolkata. Before 1905, the Surveyor General's Office was located here.

Presence in Schools and Educational Institutions

Most of the geography books at secondary levels affiliated to different boards have textbooks with a chapter on *Map Reading*. Teaching of topo-graphical maps has been one of the important parts of such courses. This becomes the first occasion for the students to know about the importance of such maps. The teachers or tutors capable of explaining topographical maps have been much in demand. There are other materials and activities organised for the students as well. One was the *Discover the Hidden Treasures of India with SOI Maps* (Fig 10.2). It contains sample topographical maps at different scales and also examples of tourist and city maps at larger scales.

Figure 10.2 *Maps for the students*

Attempts were made to make topographical maps familiar to the students. A treasure hunt was organised in New Delhi where school students were provided with maps and plane tables to locate the treasure spot (Fig 10.3). This experiment helped young students to understand the importance of topographical maps.

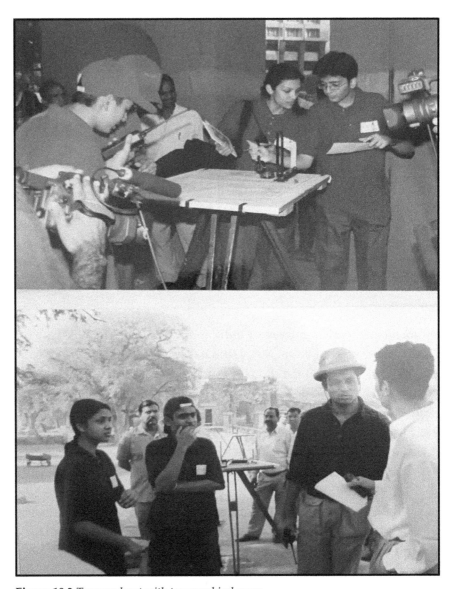

Figure 10.3 *Treasure hunt with topographical maps*

For school children, SOI has been publishing the *School Atlas*. The content includes planetary movement, maps of different continents and detail maps of India. It has also small-scale thematic maps as well. Though this atlas is meant for children, it is being used as a reference material for boundaries, placenames and other topographical details (Fig 10.4A).

Figure 10.4 *(A&B) School atlas & Aerial photographs*

As mentioned earlier, Survey of India through its Directorate of Survey (Air) located at R.K. Puram, New Delhi has been responsible for the aerial photographs in the country. It maintains the repository of all aerial photographs and meets the needs of Air Force. Since the interpretation of such photographs and photogrammetry had been included in the undergraduate and post-graduate courses in Indian universities, there has been demand of aerial photographs. Hence, some aerial photographs have been identified for educational purposes. A list of such photographs has also published (Fig 10.4B). This list contains sample photographs of different parts of the country. Nevertheless, the educational institutions are supposed to preserve them and keep in safe custody. Every year institutions must provide a certificate of *safe custody*.

SOI association with educational institutions were in other forms as well. Earlier B. Tech. courses were conducted in collaboration with the Jawaharlal Nehru Technological University (JNTU), Hyderabad. In 2003, M. Tech. courses were performed in its Bangalore premises with affiliation from the Visveswaraya Technological University (VTU). Unfortunately, both the courses did not survive longer. Further, SOI also helped the Department of Geography of Osmania University, Hyderabad to establish its GIS lab in 2002.

Presence in the Media

Since the survey operations were primarily for army and security forces, SOI was traditionally shy of publicising its activities in the media. Nevertheless, the first stamps printed in India were in SOI press in Kolkata. In 2004, a First Day Cover was released in memory of two hundred years of the Great Trigonometrical Survey and contribution of Nain Singh and Radha Nath Sikdar (Fig 10.5).

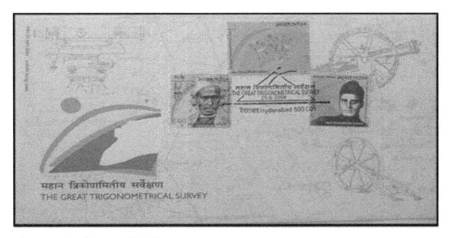

Fig 10.5 *First day cover (B/W)*

For the same reason, a wide publicity was made in the press in India and abroad. The *Frontline* magazine, apart from the other articles, brought out the following item in its issue published in July 2003 (Fig 10.6A). Further, British High Commission in New Delhi issued a *News Release* about celebrations of two centuries of mapping of India to be held in different cities of the United Kingdom (Fig 10.6B).

Figure 10.6 *(A&B) Published in Frontline & News release from British High Commission, New Delhi*

SOI rarely participated in press conferences. Such activities were associated with some events which were again uncommon. Hence, occasion to contact with common people were in connection with release of maps, participation Science Congress, establishment of new GDCs in different states of the country and the like. The examples of such have been depicted in Fig 10.7.

Figure 10.7 *SOI in press and publicity A – Raj Bhavan, Shimla; B – Science Congress, Bangalore, C – Press conference, New Delhi; D – Map release ceremony*

Participation in the Republic Day function

In 2003, on 26th January, SOI participated in the Republic Day parade in New Delhi. The booklet published on this occasion mentions that (Fig 10.8):

"The origin of Survey of India, the premier mapping agency of the Nation dated back to 1767. It has now grown into a giant organization providing the geospatial needs of the country.

The tableau of Survey of India presents the Great Arc of the meridian – the largest measurement of the earth's surface ever to have been attempted. The Great Arc made possible the accurate mapping of the country. Survey of India is now a pivotal information provider for varied applications viz. disaster preparedness, natural resources management, infrastructure development, urban development, industrial and business geographic. India is amongst the few countries of the World having the complete mapping cover of its landmass."

This was first time Survey of India participated in a Republic Day function in New Delhi.

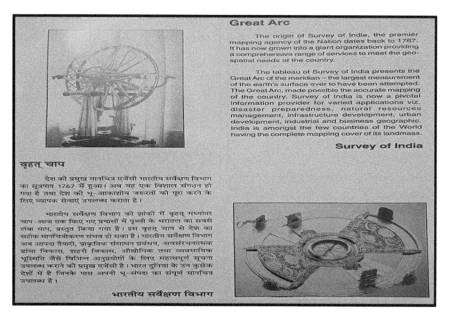

Figure 10.8 *Participation in Republic Day parade*

Presence at the Borders

Survey of India has been responsible for demarcation the borders of India. It has been working closely with the Ministry of External Affairs. Earlier it uses to have a Boundary Cell in New Delhi but under the direct control of the Surveyor General's Office (SGO). Now this office has been upgraded to International Boundary Directorate. Later a Digital Cell dedicated to Ministry of External Affairs for boundary work was established in the Survey's premises in Palam, New Delhi. It was inaugurated by Shri Kunwar Sibal, Foreign Secretary on 28th January 2003 (Fig 10.9).

Figure 10.9 *Induration of Digital Cell (MEA) in Palam*

The Surveyor General of India has been the adviser to the Government of India on all survey matters including demarcation of the external boundaries of India, their depiction on maps published in the country and also advise on the demarcation of inter-state boundaries. The Boundary Directorate has been the nodal point for co-ordination of all cartographic and topographic activities relating to the international boundaries and provides cartographic support to Ministry of External Affairs in this matter. The boundary demarcation includes following activities:

(a) To check/provide co-ordinates for newly constructed pillars.

(b) Strip Mapping.

(c) Relocation of Missing Boundary Pillars.

(d) Preparation of Special Maps required by the MEA for boundary negotiations and studies.

(e) Mapping for neighbouring countries as and when projected by MEA.

Further, Surveyor General of India has been the leader of the following committees constituted by the Ministry of External Affairs:

(a) Joint Technical Level India-Nepal Boundary Committee.

(b) Delegation to Indo-Pakistan Talks relating to delineation of boundary in Sir Creek Area and delimitation of maritime boundary.

(c) Technical level India-Bhutan boundary committee.

(d) Technical collaboration and technology transfer to neighbouring countries *viz* Bhutan, Myanmar.

India has 15106.7 km of land border which was then running through 92 districts in 17 states and a coastline of 7,516.6 km (mainland & islands) touching 13 states and union territories (UTs). Barring Madhya Pradesh, Chhattisgarh, Jharkhand, Delhi, Haryana and Telangana, all other states in the country have one or more international borders or a coastline and can be regarded as frontline states from the point of view of border or coastal management. Nevertheless, India's longest border has been with Bangladesh which include 155 km *fluid boundary* as well; while the shortest border is with Afghanistan (Table 10.1).

International borders	Sectors	Length in km
India-Pakistan	Gujarat, Rajasthan, Punjab, Ladakh, Jammu & Kashmir	3,323
India-Afghanistan	Jammu & Kashmir	106
India-China	Jammu & Kashmir, Himachal Pradesh, Uttarakhand, Sikkim, Arunachal Pradesh, Ladakh	3,488
India-Nepal	Uttarakhand, Uttar Pradesh, Bihar, West Bengal, Sikkim	1,751
India-Bhutan	Assam, West Bengal, Sikkim	699
India-Myanmar	Arunachal Pradesh, Nagaland, Manipur, Mizoram	1,643
India-Bangladesh	West Bengal, Assam, Meghalaya, Tripura, Mizoram	4,096.7*

Table 10.1 *India's international borders*

Note: *Excluding length of enclaves of India and Bangladesh*

The neighbouring countries are Pakistan, Afghanistan, China, Nepal, Bhutan, Myanmar, and Bangladesh. The India-China border has three parts (a) Eastern Sector (Arunachal Pradesh), (b) Middle Sector (Sikkim), and (c) Western Sector (Undivided Jammu & Kashmir, Himachal Pradesh, and Uttarakhand). For these three sectors negotiations are in progress.

Figure 10.10 *(A) Joint survey meeting at India (West Bengal)-Bhutan border; (B) India-Myanmar Meeting regarding joint inspection of the border etc.*

The India-Nepal border of 1,751 km is primarily based on the Treaty of Segauli of 1815. It is an open border, and certain pockets have *riverine segments*. Further, the India-Bhutan sector is of 699 km was settled in pursuance of Treaty between British India and Bhutan which was concluded at Sinchula in November 1865 (Fig 10.10A). This is also an open border but have limited number of entry points in comparison to Nepal. Furthermore in the east, the India-Myanmar border of 1,643 km is based on boundary agreement between Republic of India and the then Union of Burma signed at Rangoon in March 1967. Demarcation at a few places is yet to be completed. The trijunction between India, Myanmar and China can only be fixed after a tripartite agreement (Fig 11.10B).

The India-Bangladesh border of 4,096.7 km has been settled by the following bilateral treaties and boundary awards:

1. Radcliffe Award, 1947

2. Bagge Award, 1950

3. Nehru-Noon Agreement, 1958

4. Indira-Mujib Agreement, 1974

5. Exchange of enclaves (*Chitmahals*) Agreement, 2015.

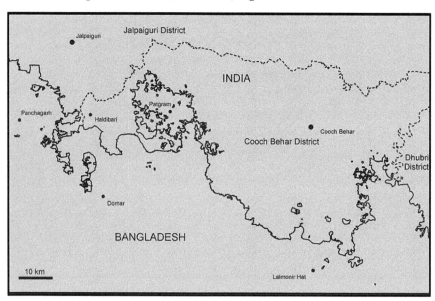

Figure 10.11 *Nature of India-Bangladesh border.*

In the agreement of 2015, there was an exchange of 111 Indian enclaves (17,160.63 acres) and 51 Bangladeshi enclaves (7,110.02 acres) (Fig 11.11). Along with this agreement, the issue of adverse possession and undermarketed borders were also sorted out. The involvement of SOI in maintaining India-Pakistan border of 3,323 km is more due to historical reasons. There is a *Line of Control* indicating the illegal occupation of a part of Jammu & Kashmir by Pakistan. Following Katchchh Accord, part of this border (Pakistan-Gujarat) was established in 1965. However, Sir Creek area is yet to be settled. It has a 96 km tidal estuary bordering India (Gujarat) and Pakistan (Sindh) and has an outlet in the northern Arabian Sea. It is located at approximately 23°58'N 68°48'E in uninhabited marshy lands of the Indus delta, just to the west of the Great Rann of Katchchh. During the monsoons (June and September), this area is flooded and envelops the low-lying salty mudflats around it. The settlement of this sector has also implications in demarcating the territorial sea and the exclusive economic zone (EEZ). Furthermore, the Indo-Afghanistan Border is only of 106 km having boundary with the Jammu & Kashmir state. As it is evident from the above, each international boundary has various issues attached with them. The Indian surveyors have to understand the historical, geographical and geopolitical realities in order to act on the ground.

In addition to land boundary, SOI has been maintaining the coastline which is monitored through different stations located along the coast. Continuous observation is important because coastline determines the heights of different places in the country. Further, it facilitates in understanding coastal ecology and helps in disaster management, including tsunami. In some places, the issues related to the coastlines and the land boundaries are to be considered together. Examples are of Gujarat bordering with Pakistan and West Bengal with that of Bangladesh.

The international borders are continuously being monitored, and the boundary pillars are repeatedly visited. Further, the International Boundary Division of the Survey of India, in collaboration with the Ministry of External Affairs, also carries out detailed study of each sector of the international borders with the help of maps, agreements, historical documents, gazetteers, geographical situations, satellite imagery and GPS.

Cooperation with the Ministries of External Affairs and Defence

Apart from maintenance of international boundaries, Survey of India also provides cartographic support the Ministry of External Affairs (MEA). As mentioned earlier, a Digital Cell was established in the Palam office premises for preservation of historical records (Fig 10.12). In addition, SOI has also assisted to develop a web site to access information regarding Indian embassies, high commissions and other offices abroad (Fig 10.12).

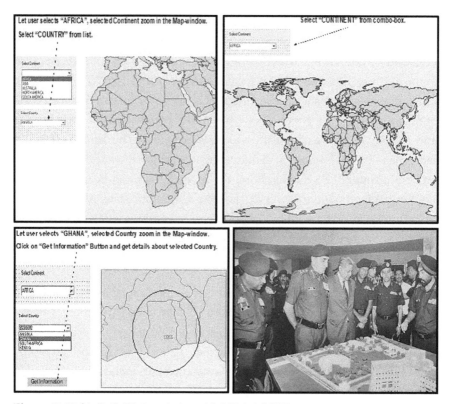

Figure 10.12 *(A, B. C, D) Association with MEA & MOD*

From very inception, SOI had its roots in the army. Either under the East India Company or directly under the Crown or even after independence, the surveyor generals and other senior offices were from the army. The topographical maps were basically for the armed and security forces. SOI also prepared maps and navigation charts for Air Force. The Directorate of Survey (Air) was meant to have an interface with this Force. Further, army officers have been on deputation and reverse deputation from and to SOI. Even the cadre controlling authority for army officers has been the Surveyor General of India who may or may not be an army officer. Regarding technical work, there has been a close cooperation between the SOI of the Ministry of Defence (Fig 10.12D).

International Linkages

The activities of Survey of India were not only limited to undivided India. We do not reference about Survey officers involved in different countries or in demarcation of boundaries. Some adventurous journeys were carried through Afghanistan and Turkistan by Alexander Burnes in 1932 during which he took assistance of two Indians, Mohammad Ali, an engineer from Bombay and Mohan Lal. The latter was appointed as Munshi-writer on 18-12-1831 from Delhi. He maintained a detailed diary of the journey through Kabul, Bukhara and Mashed and the return through Herat, Kandahar and Peshwa. His fascinating journal was published in Calcutta in 1834 and gave a vivid account of Turkistan and its people. The first expedition was undertaken by one Muhamed-i-Hamed who went from Ladakh to Yarkand by the Karakoram Pass and fixed the latitude of Yarkand. Earlier in 1914, survey officers helped in demarcating Turco-Persian Boundary in collaboration with Russians, Persians and Turks. Other group was sent for scientific exploration and mapping the headwaters of River Yarkand. Further, SOI extended survey support to Indonesia, Nigeria, Bhutan, Nepal and Iraq. It has also undertaken various jobs of demarcation of international boundaries *viz*. Bangladesh, Pakistan, Burma and Nepal.

Apart from the survey-related activities, SOI has been taking part in different training and scientific activities around the globe. Some of the examples of such participations have been mentioned below:

(a) Publication of *Indian Tidal Tables* for port around the Indian Ocean (Chapter 6).

(b) Participation in Scientific Expeditions to Antarctica Programme to generate Geodetic & Geophysical data for scientific studies and Mapping of Maitri & environs.

(c) Participation in Technical Cooperation programmes *viz*. Undertaking development surveys, upgradation of surveying technology in the field of geodesy, digital cartography, digital photogrammetry and training of officers for the Survey Departments of neighbouring countries.

(d) Member of International Cartographic Association.

(e) Member of Executive Board of Permanent Committee on GIS Infrastructure for Asia and the Pacific (PCGIAP).

(f) Member of Working Groups on Cadastre and Institutional Strengthening under the United Nationa Regional Cartographic Congerence for Asia and the Pacific (UNRCC-AP).

(g) Participation in Cambridge Conference.

At the global level the presence of Survey of India had been felt for different reasons. Ordnance Survey of United Kingdom had appreciated its role in the Cambridge Conference in 2003. Surveyor General was invited to join its Advisory Committee and to look after the interest of Asia (Fig 10.13 A, B & C).

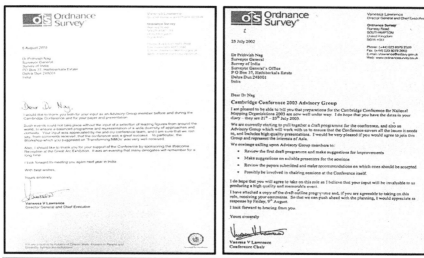

Figure 10.13 *(A, B & C) Appreciation at Cambridge Conference*

Along with the Cambridge Conference during July 2003, an exhibition of SOI was organized in London, United Kingdom. This event was held on the occasion of 200th year of the *Great Trigonometrical Survey*. This event was attended by the Union Minister for Science & Technology, Government of India, Prof Murli Manohar Joshi and Secretary, Department of Science & Technology, Prof V.S. Ramamurthy and several other officers from the

Survey. Indian High Commissioner and several other persons were present on this occasion (Fig 10.14).

Figure 10.14 *Survey of India in London Exhibitions, 1862 & 2003*

SOI efforts were also appreciated in the global events organized under the banner of the Permanent Committee on GIS Infrastructure for Asia and the Pacific (PCGIAP). Apart from its conference in Phillipines, SOI organised its conference in Mysore just prior to the Global Spatial Data Infrastructure congress held in Bangalore in January 2004 (Fig 10.14). Earlier, SOI participated in Executive Board Meeting of Permanent Committee on GIS Infrastructure for Asia and the Pacific (PCGIAP), 28-30 October 2002 in Makati City in Philippines. It also organised an International Workshop on Institutional Strengthening as Chairman of the Working Group 4. Further, its contribution was appreciated by the PCGIAP as well (Fig 10.15).

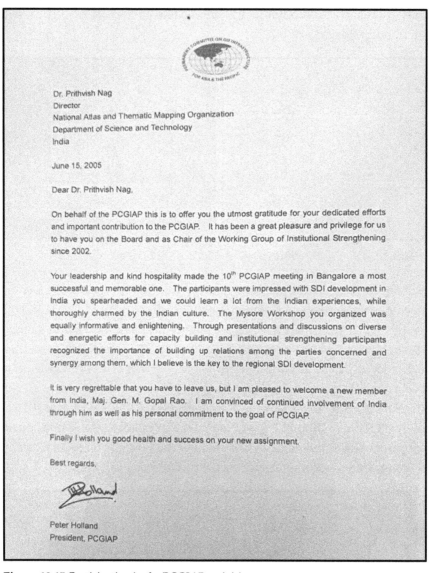

Figure 10.15 *Participation in the PCGIAP activities.*

Furthermore, the International Institute for Geo-Information Science and Earth Observation, popularly known as ITC, invited the Surveyor General of India to deliver a lecture on "The Indian Experience in Developing a National Spatial Data Infrastructure" in a seminar on *The Future of Geospatial Data Infrastructures* organized in honour of the contributions and marking the retirement of Prof. Ir. Richard (Dick) Groot (Fig 10.16). It was a rare occasion when the Surveyor General is invited of scientific deliberations abroad.

Dr. Pritvish Nag
Surveyor General, Survey of India
The Indian experience in developing a National Spatial Data Infrastructure

Technological advancements in the fields of computers, satellite sensors, global positioning systems, geographical information systems and digital photogrammetry along with increased pressure to obtain, analyse and apply spatial data have forced even a developing country like India to think of evolving mechanisms to make both spatial and non-spatial data accessible to users. India has possibility the longest tradition of systematically collected scientific spatial data, Survey of India having been doing so for 235 years.

It has been felt in the last two years that the lack of metadata was seen as the major problem. Secondly, the resources required to convert the analogue data into digital form made this a stupendous task. Further, the challenge lies in bringing round the various data producers such as surveyors, geologists, remote-sensing scientists, foresters and hydrologists to think alike. They all are possessive of the data and had initially expressed reservations in sharing their data assets. Another related issue is conversion of the analogue data following prescribed standards and network protocols. These issues are yet to be fully settled.

India has been classifying the topographical data that comes in the way of making it available over the net. One of the proposed solutions is to bring out a new series of maps exclusively for civilian use. However, the content is yet to be decided. While these problems are still to be resolved, India is going ahead with the NSDI.

Figure 10.16 *Surveyor General at ITC, The Netherlands.*

Establishment of National Geotechnical Facility, Dehra Dun

Following a scientific visit by a DST delegation headed by the Surveyor General of India in September 2008, the Ministry of Science & Technology approved the establishment of a National Geotechnical Facility (NGF) in Dehra Dun with technical cooperation from the Norwegian Geotechnical Institute (NGI). It was recommended that NGF becomes an organisation offering both engineering and design capability and testing facilities; research and consulting; investment in international cooperation and outreach, and in training of its personnel; and to give priority to the following competence areas: natural hazards, sustainable infrastructure, computational geomechanics (also offshore applications), laboratory and *in situ* testing, and monitoring, remote sensing and early warning. Further, it was recommended that NGF should become an autonomous research and consulting organisation within the Ministry of Science and Technology. The vision was as follows:

"NGF should aim at becoming within the first five years the premier non-university geotechnical organisation in India, and an internationally recognized authority in selected fields of expertise. NGF should be a competence centre providing both research and consulting expertise, with cost-effective solutions to geoproblems, improved and safer infrastructure in India and safe exploitation of natural resources, all objectives to benefit India's society."

This facility was established in the Survey of India office (6 DO Party, Poonch House, Circular Road, Dehra Dun) having an area of about 4.5 acres of land and two buildings. Apart from the SOI, DST entrusted the functioning of this facility to Indian Institute of Technology, Roorkee and the Wadia Institute of Himalayan Geology, Dehra Dun. Several equipment was provided to develop the infrastructure. The geospatial facilities included total station, differential GPS, remote sensing and GIS software like ArcInfo, Erdas, TNTMips and MapInfo.

Conclusion

Apart from the maps, atlas, aerial photographs and certified maps in books, the salient presence of SOI has been experienced when maps and now geospatial data are required for the construction of dams, railways, roads, townships, canals and the like. Its maps find their way to the Library of Congress in Washington, New York Public Library, British Library, National Library (Kolkata), universities, survey institutions and earth science departments. Indian Army and Air Force have been its eternal patrons.

▼ *This portrait of Col William Lambton (1753-1823) has now been completely restored*

▲ *This portrait of Col William Lambton (1753-1823) has now been completely restored (renovated at London as Phillimore's Records of Survey of India, vol. III, p. 467)*

Figure 10. 17 *Col Mackenzie & Col William Lambton*

Unfortunately, there is no budget for the publicity of its products and services. Hence, the users come to know about the survey products silently but surely as no institution provides maps and geospatial data of highest level of accuracy. Its efforts have been appreciated after decades when the actual actors were no longer around. Moreover, such appreciations have been by a very few admirers only. No other department of the Government of India suffered in a such a way though contributing for centuries in a very critical sector of development. Nevertheless, the contributions of SOI Officers like Colonel Colin Mackenzie and Colonel William Lambton have been appreciated all over (Fig 10.17). But, in the country, Survey of India has yet to get its right gratitude.

Chapter 11

Retrospection & Foregleam

Collection and depiction of information has been very important for analysis for planning purposes. Map forms an important medium of representation of spatial information. In India, spatial data represented on maps have been collected keeping in view of the military use and security of the nation: detection, preparedness, prevention, protection, response and recovery. The geospatial community had realized the fact that these data have utility far beyond the of national security. Disaster management, community development, developing a successful society, management of natural resources, health care, economy and support to scientific community for research requiring spatial data. Development of spatial data in these areas are pursued by the public, private and non- governmental agencies and individuals.

Sustainable Spatial Data Infrastructure

Efforts of SOI, the national mapping agency, to provide national coverage, had been commendable. The analogue data on 1:50,000 scale, was converted into digital form, as a committed programme, in a short time. The demands of the society were to have more current and accurate geospatial data. This requirement cannot be met from the existing digital spatial data. Hence there was a need to avail the opportunities provided by new technologies to collect, transform, manage, disseminate the spatial data. These opportunities provided avenues for participation by more individuals and organizations to undertake these tasks in a more flexible policy framework and to meet more vibrant demands of geospatial data. Most of the requirements were common, thus providing opportunity for partnerships among the central and state government agencies, public bodies, non-governmental agencies and individuals. Private sector investment in the development and maintenance of the data, provided an opportunity to work together with those agencies.

New technologies in the process of creation, development, maintenance and dissemination of a spatial data Infrastructure, requirement of accurate and current data and opportunity for partnerships were the components of the spatial data infrastructure. Proper harmony between all these components was required to ensure that they consider the individual needs and at the same time meeting the broader needs of the society.

The national mapping agency should provide the basic framework spatial data which is accurate and current, enabling development and implementation of applications by participating agencies, at various levels, without unnecessarily competing with them. An operational approach should be adopted by such an agency, that takes advantage of partnership among the network of organisations, to acquire, develop, and access the spatial data; and at the same time, to practice an organizational and business approach maintaining a leadership role in data management. In the emerging trends, national mapping agency should be performing the role as:

(a) A guarantor of consistent, up-to-date, accurate and complete spatial data.

(b) An organizer responsible for awareness, availability and utility of the spatial data.

(c) A catalyst and collaborator for creating and stimulating partnerships.

(d) A certifier and integrator of spatial data generated by participants.

(e) Data owner, producer and solution provider when need arises.

(f) A leader in development of standards for geo-spatial data.

Status of Spatial Data and Products

Interaction with the users has been necessary regarding understanding of the product requirement, acquisition of data by different techniques, transformation of collected data to produce desired products, and finally delivery of the same. Dissemination of information in time to meet the user needs, management and archival of the data for quick retrieval with proper security were the basic processes in spatial data generation and use.

Data collection is a very important stage in the process. The precision of information and its truthful depiction had been the primary requirement. Nevertheless, the accuracy of information depends on the purpose and scale of depiction. Providing a temporally up to date data to the user is the responsibility of the data provider. In the field, instrument and method-ology have to be considered for data collection which includes:

- Triangulation of different standards to be used for establishing framework in the area of survey.

- Traversing has to be carried out depending on the terrain condi-tions using theodolite for angle measurement and chain for linear measurement.

- Collection of height data using leveling instruments.

Earlier, for detailed surveys, plane tabling was employed, in which inter-section, radiation and resection techniques were used for determining the location details of information. All these methods were time consuming, and lot of inaccuracies were creeping in due to the methods employed, besides getting temporally outdated by the time the data is supplied to the user. They had following limitations.

(i)   Inter visibility of the ground points

(ii) Accessibility of the points

(iii) Possibility to occupy the station and make observations.

Aerial photographs were used since 1960's for data collection using analogue machines and generating hard copy outputs. Satellite imagery could be used mainly for land use assessment. Resolution of the satellite imagery, till recently, was not good enough, for capture of topographical features and meeting the mapping standards. Products were sometimes generated without scientific assessment of user requirements, mostly by adopting manual techniques. The computer aided cartographic technology was being used in a limited way. Thus the digital technology could not be wholly exploited in generation of products and product multiplication. The use of the spatial data was mainly in the form of interpretation from hard copy output.

Changing Needs

The then practices and attitude could not meet needs of the user and accomplish the role expected from a national mapping agency. It was to make the spatial data available in the most appropriate format:

(a) Geodetic information and services.

(b) Topographic information and services.

(c) Land information management and services.

(d) Information management, quality product generation and on time delivery.

For proper integration of spatial data produced by different agencies at different levels, there should have been framework data with proper connectivity and should be able to be used by all the stake holders. On this framework data, thematic information could also be depicted for different applications.

Spatial Data Transformation & Management

Transformation of spatial data required editing, integration of data produced by different sources, and application for product generation as per user needs. The trend had so far been to print the hard copy map for providing to the user. In the digital environment, it is easy to handle data in digital mode rather than hard copy. Printing of maps however cannot be stopped altogether, and use of data in hard copy form will continue for some time. Four colour printing and reproduction can be accomplished by taking advantage of geospatial technology. The entire maps on 1:50,000 scale are now in digital form. A range of cartographic and other data products could be generated from this data base in a user-friendly format. SOI was to be engaged in developing following digital and hard copy products:

   i. City maps, guide maps, route maps etc.

  ii. Image maps using DTM from existing maps.

 iii. Layers of information as required by users.

iv. Tourist maps, trekking Maps, district planning maps etc.

v. Develop data base from large scale aerial photographs for urban planning and rural development.

vi. Information KIOSKS at important places for awareness of public.

vii. Application tools for users in GIS environment for interpretation, analysis and retrieval assisting in the decision-making process at various levels.

Providing accurate and current data has been the responsibility of the national mapping agency. We have to device methods of data structure and storage by organising the highly complex data in layers so that it is live, consistent, accurate and flexible. SOI was to establish national centres to act as nodes of NSDI. All state geospatial data centres were to be interconnected in order to make the data available on demand to the stake holders in most user-friendly environment.

Spatial Data Dissemination

The technology is of lesser relevance if the data is not made available timely for solving problems. Crisis management and recovery has been an important issue in mapping operations and geospatial industry, with emphasis on real time access and retrieval for critical information. Making data so available to meet the user needs has been of primary importance at local, regional, national and global levels. Every day spatial information makes the life easier and simpler. We have to identify methods to provide the products generated in a user-friendly manner, both electronically and in traditional hard copy form. Accurate web enabled spatial data in the following areas needs to be made available on minute-to-minute basis:

i. Weather conditions.

ii. Traffic and road conditions in emergency.

iii. Tracking of freight movements

iv. Spread of disease.

v. Data from satellite imagery for national security.

Open Architecture

Technology trends have seen web enabled services emerging with the demand for quick access to enterprise data, with interoperability and standards playing an important role in serving data regardless of the format. Standardizing data formats and working out exchange formats may no more be a priority. With interoperability, users will no more be captive to proprietary formats or GIS applications because data servers can access and manipulate data in its native format. Legacy data no longer requires migration, translation and conversion. The power of open architecture allows users to bye-pass these functions. Sophisticated N-tier architecture allows organizations to provide a variety of data and access to groups across the enterprise, via an intranet or internet, facilitating live connections to real time information.

Interoperability has been a growing trend. The Open GIS Consortium (OGC) and GSDI were still carving out standards-based specifications and promoting a national data infrastructure in which sharing of information was to be easy and in real time. Technological capabilities must meet human needs. The geospatial community is expected to work together to remove barriers and achieve success.

Geodetic Control

The *Great Trigonometrical Survey of India* in its over two and half centuries, collected valuable information for the service to the humanity and science. Its products and wealth of information are being utilized in the development of the nation. Geodetic and geophysical information like planimetric and height control, gravity, geomagnetic and sea level data were collected with an aim of serving humanity by forewarning them for the natural disasters and by bringing the information of resource availability to the earth scientists.

Timely and accurate information on natural resources has been prerequisite for their optimum utilization and effective management. The importance of collecting, using and updating the geospatial data on resources covering areas of interest has been emphasized from time to time. It is important to indicate that how and where the resources are located in an area and how the various resources are spatially interrelated. Cartographic information in the form of maps, charts, atlases, digital data, GIS and other form of geospatial data are the prime requirements for effective management of resources such as water, land, forest and minerals.

SOI provides national level maps and GIS to support various activities related to expeditious and integrated development so that all resources contribute their full measure to the progress, prosperity and security of the country. Maps on various scales are being prepared in the department which show feature like natural, man-made and elevation in the form of contours etc. These maps could be of immense help in water management and developmental activities. GIS along with remote sensing, GPS, aerial photography, laser mapping and land survey methods are now considered as the essential components of the geospatial technologies.

In a developing country like India, having huge population base with democracy, public access to information and freedom of expression, geospatial technologies have several additional functions to play. It is considered as a tool for getting in touch with the government, lodging protests, and having interface with people. In addition, there is a wide scope for business enterprise as well. Geospatial technologies have a definite role to play in the following sectors:

    (i) Highways survey and mapping.

    (ii) Natural disaster management.

    (iii) Telecommunication services.

    (iv) Power projects.

    (v) Ports projects.

    (vi) Urban land use planning.

    (vii) Rural land management or agricultural land management.

    (viii) Waste land reclamation projects

    (ix) River basins or watersheds management.

(x) Imagination and canal developments.

(xi) Consolidation of land holdings and land records.

(xii) Digitization of maps on various scales.

(xiii) Updating and digitization of wards and services maps of different cities and towns for their betterment both in terms of economy or finance and environments.

(xiv) Mapping of places of tourist interest or pilgrimages both in hard copy and soft copy forms.

(xv) National natural resource management.

(xvi) Monitoring of environment

(xvii) Terrain and height information or positions for frequency requirement for the purpose of All India Radio, *Door Darshan* (television) and telecommunication

(xviii) For better and fool-proof security with up-to-date and accurate maps in the form of digitized or special software both for military and civilian purposes.

(xix) Development of accurate navigational maps both at ocean or water and land for better traffic-flow and least risk of accident.

(xx) Production of good quality maps and atlases for the use of students and others.

(xxi) Maps and GIS for government officials and the common people to understand the real positions and facts in their day-to-day jobs or life.

Indian Geospatial Industry

It is seen that the Indian industry is fully capable and able to take the challenges of the demand for mapping or cartographic work. The concerned industry is entirely prepared to take cartographic work in collaboration with the government departments or agencies. They have been doing a number of cartographic or mapping projects for the government. They are also involved in GIS or Land Information System (LIS), digitization of maps or documents etc for different organisations and agencies. It was also found that there has been a great demand for thematic maps on large and small scales on 1:1,000, 1:10,000, 1:50,000 and 1:1,000,000 for projects in the field of telecommunication, business, irrigation, power, transport and tourism, resource management, town and urban planning and the like.

It has been assessed that there is a huge potential for export of cartographic products and services, especially in the field of GIS or land information system, scanning and digitization, remote sensing technology and developing different products, services and applications not only for India but other countries as well. Many Indian industries have done mapping and digitization projects for foreign companies or countries and are doing for many others. Further, several Indian industries are ready to collaborate in the field of geospatial technologies with global partners. Furthermore, India has become a hub for attracting IT projects from abroad. Mini-Bangalores have cropped not only in the fringes of the bigger Indian cities, but the medium size cities have not been spared as well. With the successful experience of IT, trained man-power available with fairly good knowledge of English and capacity to pick up other foreign languages, India is likely to become hub for global GIS or geospatial industry as well. Geospatial institutes for teaching, training and research are coming up in every city. The time is not too far when India will become a favorite destination for geospatial industry as well. Geotechnology is the new challenge - a determination to promote geospatial technology is considered to be a new initiative in academia today as it has become one of the fastest growing high-tech careers for students. It provides powerful tools for geographic analysis for almost any academic discipline. GIS allows students and researchers to ask and answer geographic questions by designing and analyzing maps using user-selected criteria. It is widely used for administration, research, community services, libraries, community colleges and teaching. GIS is multi-facet and hence a wide range of applications.

Trends

The economic growth, market forces, government policies and technological advancements have changed the basic understanding of map-making.
The challenges are enormous, and as it happens in a developing economy,
when we try to grip one basic issue, two other get out of hands. However,
other developing countries do look forward to India about our experiences.
Nevertheless, the actions in SOI have been making ripples all over. Hence,
this organisation must transform even at the cost of much internal resistance.
Now every state has a geospatial data centre. These centres are meant to take
care of the requirements of each state, their governments and their development agencies. *Data for Development* has become the new *mantra*. Further, the
focus has shifted from printed maps to digital data, hence *Geospatial Data
Centres* replacing typical directorates have been a part of the transformation
process. To co-ordinate these data centres, the National Geospatial Data
Centre or NGDC has been established in Dehra Dun. For printed maps, the
printing establishments have been decentralized. Furthermore, considering
the voluminous demand of applications, a new directorate for GIS and
Remote Sensing has been established in Hyderabad.

The G&RB has been facelifted to meet the new and emergent challenges like
tsunami. A major activity of this directorate was to establish new control
points tuned for WGS 84 series of maps. Survey Training Institute or STI
has introduced new courses on GIS and NSDI. An international course on
SDI was organized under the aegis of PCGIAP in Hyderabad.

The new *National Map Policy* was approved by the government in 2005. The
Committee of Secretaries has approved the policy the way it was proposed.
Nevertheless, we faced a lot of resistance in bringing it out. There were
people within the organization and outside who were interested in
continuing the old restriction policy (Chapter 5).

National Expectations

India is fast moving into information and knowledge society. Emphasis is
increasingly placed on IT driven "transparent" e-governance. The nation
has been generating voluminous field and digital data *i.e.* information
through systematic topographical, geological, soil and cadastral surveys.

Access and availability of such information to the citizen, society, private enterprises and government are of immense importance. As a part of this vision, NSDI is being involved through partnership approach among various data generating agencies to facilitate integration, easy access and networking of databases. NSDI has been conceived as a national system that synergistically combines the resources and infrastructure of various players with the power of IT and enabling information support for decision making in government, industry, academia and other organisations besides serving the public needs. NSDI strategy and action plans have been formulated through multi-institutional approach involving DST, SOI, ISRO, GSI and NATMO. Hence, keeping in view of the potential, expertise and expectations, then a new vision for SOI was proposed.

Keeping the above objectives in mind, this nearly 250 years old institution was reorganised. Since then, there has been a direct communication between the Surveyor General's office (SGO) and the field units. Further, in order to take care of the development needs of each state of the country, restructuring was carried out. The specialized directorates have been given fresh mandates as well. The new arrangements were difficult to implement, and the associations were resisting such changes. Moreover, the new arrangements forced the SGO to change as well. There is an increased workload for officers in dealing with 40 directorates well spread over the country. Technical issues have been included in the functioning of SGO which was otherwise having purely administrative responsibilities.

Implications of the National Geospatial policy 2022

For past several years there has been attempts to bring out a national geospatial policy keeping in view of the fast-changing technology, new data sources, global scenario and possible role of geospatial data in every domain of the Indian economy. Locational aspects have been considered pivotal for developing strategies and providing services to common people of the country. As a result, *Guidelines for Acquiring and Producing Geospatial Data and Geospatial Data Services including Maps* was issued on 15 February 2021. Following these guidelines, a *National Geospatial Policy (NGP 2022)* was announced in December 2022. This policy suggested certain changes in the structure and functioning of SOI.

In the policy, an ecosystem is to be developed for innovation as well. According to the National Innovation Act 2008: "Innovation means a process, which may be break through or incremental, by which varying degrees of measurable value enhancement is planned and achieved, which may occur systematically or sporadically in any commercial activity." Hence, there is a commercial angle to geospatial data. Further, the NGP 2022 is to be considered with the Innovation Act to (a) facilitate public, private or public private initiatives; (b) improve the environment to support innovation; (c) facilitate building an innovation eco system; (d) evolve an integrated science and technology plan; and (e) provide for law of confidentiality. Hence, SOI is supposed to have a proactive role in this regard. Its geospatial, GTS and CORS data is to be considered as a *public good*. Earlier attempt in 2012 in the form of National Data Sharing Policy is to be fully implemented.

The objective is to be a partner of the thriving geospatial industry in the country involving private enterprise. For this purpose, milestones have been proposed specifically for the years 2025, 2030 and 2035. Apart from liberalization, commercial and democratization of geospatial data, the onus on SOI is to redefine the *National Geodetic Framework* using modern positioning technologies and provision of online access; and to provide high accuracy geoid for the entire country. It also expected to take part in the production of:

(a) High resolution topographical survey and mapping (5-10 cm for urban & rural areas and 50 cm-100 cm for forests and wastelands).

(b) High accuracy Digital Elevation Model (DEM) for entire country (25 cm for plain, 1-3 metre for hilly and mountainous areas).

(c) Develop a Geospatial Knowledge Infrastructure (GKI) underpinned by Integrated Data and Information Framework.

The policy also expects high resolution/accuracy bathymetric geospatial data of inland waters and sea surface topography of shallow/deep seas in order to support *blue economy*. The other expectations are (a) survey and mapping of sub-surface infrastructure in major cities and towns; and (b) development of national digital twin of major cities and towns. The overall focus is to make geospatial technology and data as agents of transformation for achieving the *Sustainable Development Goals* (SDGs), bringing efficiency in all sectors of economy and instilling accountability and transparency at all levels of governance.

An institutional framework has been constituted in the form of a Geospatial Data Promotion and Development Committee (GDPDC) at the national level which shall be the apex body for formulating and implementing appropriate guidelines, strategies and programmes for promotion of activities related to geospatial sector. GDPDC shall drive the overall development of the Geospatial ecosystem replacing the earlier functions and powers of National Spatial Data Committee (NSDC). Surveyor General of India will be the Member-Secretary of this committee. Further, the composition, powers and functions of existing NSDI Executive Committee is to be realigned accordingly in order to make the NSDI mechanism more robust, efficient and effective.

SOI shall be the agency responsible for developing and operating the *National Geospatial Data Registry* (NGDR) and the Unified Geospatial Interface (UGI) in collaboration with Bhaskaracharya National Institute for Space Applications and Geoinformatics (BISAG-N), other institutions and the private sector, under the guidance and supervision of GDPDC in relation to the scope, functionality, and performance of the NGDR and the UGI. The earlier *National Map Policy 2005* (Chapter 5) is to be replaced as well. The new geospatial policy has modified the mandate of SOI. About the role and organisation, this policy of 2022 mentions that:

> "While SoI will continue to be the overarching nodal agency for Geospatial Data, only the generation/ maintenance of minimal foundational data/core functions would be performed by SoI. SoI may also involve private sector and other surveying entities such as GSI, FSI, etc.

> Amongst the 14 National Fundamental Geospatial Data Themes, Geodetic Reference Frame, Orthoimagery and Elevation are most pivotal because together they provide the Geodetic and Digital Spatial Framework that act as common reference (X,Y,Z) for the assembly and maintenance of data pertaining to all other Fundamental and Sectoral Data Themes. When interpreted, Ortho-im-

agery and Elevation act as the source for many other Fundamental and Sectoral data.

SoI would be responsible for maintaining Geodetic Reference Frame, Orthoimagery, Elevation (DEM), Functional Areas (Administrative Boundaries) and Geographical Names (Toponymy) in collaboration with various stakeholders including the private sector by suitably aligning with the priorities of the Government, while adhering to the Goals set out in the Policy.

While within the government SoI would play the lead role for maintaining high resolution /high spatial accuracy Orthoimagery, private sector will be free to take up creation, maintenance and use of such data suitable to their requirements. Department of Space will similarly play the lead role for generating Orthoimagery of high temporal accuracies using space-based technology.

For creation and maintenance of remaining National Fundamental Geospatial Data ..... Ministries/Departments would increasingly engage with private sector to meet their requirements. They will bear the cost for the creation and development of Geospatial Data required by them and they must explore procurement of Geospatial services on their own under the liberalized Geospatial regime, rather than use SoI as an intermediary.

SoI will act as facilitator in harmonization of the data sets created using public money to ensure that data generated from various mapping activities by various stakeholders get seamlessly integrated into Geodetic Reference Framework and develop a mechanism to facilitate consolidation of the data sets into the national topographic template to meet the demand of periodically updated high-resolution and accurate topographic data for the country.

The organizational structure of SoI would be aligned with the changed Geospatial data regime, with focus on facilitating and nurturing a vibrant domestic Geospatial services industry. SoI would be transformed into a fully civilian organization. Defence stream of recruitment in SoI would be discontinued and defence stream officers seconded to SoI would be permanently reverted to Military Survey, Ministry of Defence. Requirements of fast changing skill sets in SoI would be met by domain experts sourced from the market.

Apart from the responsibilities mentioned above SOI has been entrusted with many other segments of implementation of the policy, such as (a) for creating job-opportunities, (b) SOI is to work together with DST along with experts from the industry and academia with the National Skill Development Council (NSDC); (c) to develop its National Institute for Geo-informatics Science and Technology (NIGST, earlier Survey Training Institute) as centre of excellence; and (d) to promote online courses in collaboration with *iGOT Karmayogi* (Department of Personnel & Training, GoI) platform. Further, apart from the conventional responsibility of geodetic reference frame, administrative boundaries, elevation and depth and the like, the *National Fundamental Geospatial Data Themes* will include additional features in the geospatial data such as water, transport networks, land cover and land use, land parcels, addresses, geology, soils and population distribution. The latter components have been entrusted to other departments and ministries.

Conclusion

Survey of India's existence for over two and half centuries has witnessed a lot of changes in its policy, technology, man-power requirements, infrastructure and most important is it role in the development of the territory or the country and economy. Its continuity from being under the East India Company to the present independent regime has no parallel in the world. Its scientific activities, accuracies in maps and geospatial data, and record-keeping has been unique. To write a memoir about a very limited period has been a challenge as previous considerations and future challenges were considered as well. This memoir is not an exception. Nevertheless, the objective has been to cover a record of a period which was very active in terms of new policies, establishment of NSDI and NGI, change in field operations, conversion from maps from thick glass plates to digital format and finally to GIS layers, transformation from analogue to digital operations, introduction of the open series of maps, reorganization of the whole setup, do away with OC parties, plane table surveys and drawing offices, introduction of remote sensing data, experiments with ALTM, managing disasters like tsunami, participation in global events and the like. All these activities were running parallelly sometimes complementary to each other while in some cases contradictory as well. The objective was also to record the scientific and technological contributions so that they are not lost during the course of time or get overshadowed by other enterprises of that time.

Survey of India does provide an example how an old scientific department can be transformed. The objectives, technology, products and services may vary, but it continues to be known for its quality of geospatial data. Further, history will assess how the developments and challenges were accommodated during a very short span of time. If the first half of the 19th century is known for the *Great Arc initiatives,* the early years of 21st century will be known for the *Great Digital-Transformation initiatives*. In the former phase, the Indian cartographic empire was built (Nag, 2015); while in the latter, the foundation of the digital geospatial backbone was laid. The implementation of the recently announced *National Geospatial Policy 2022* is likely to take advantage of such an initiative. The integration of both the enterprises, *i.e.* geodesy and digital technology, is likely to create waves and ripples in developing digital India – a much desired future to meet the projected US five trillion-dollar economy. Perhaps, someday, one of the successive surveyor generals will take the pains to evaluate this phase of Survey of India and include his observations in similar publications.

Bibliography

Adams, T. and S. Dworkin (1997), *WfMC Work or Handbook*, chapter Work of Interoperability Between Businesses, pages 211-21. John Wiley & Sons.

Agarwal, G.C. Lt. Gen (u.d.), *A Brief History of Survey of India, 1767-1967*, Survey of India (mimio).

Agrawal, N.K. (2004), *Essentials of GPS*, Spatial Networks Pvt. Ltd., Hyderabad.

Alameh, N. (2001), Scalable and Extensible Infrastructures for Distributing Interoperable Geographic Information Services on the Internet. PhD thesis, Massachusetts Institute of Technology (MIT), Massachusetts, United States of America.

Alfred Leick, Alfred (1995), GPS Satellite Surveying, John Wiley & Sons Inc.

Anderson, Teri, Ali Roshannejad & Naser El-Sheimy (2005), MMS on a roll, *Geospatial Today*, vol.4, issue 4, pp.26-32.

ANZLIC (1996), National spatial data infrastructure for Australia and New Zealand, *ANZLIC Discussion Paper*, online: hhtp://anzlic.org.au/anzdiscu.htm.

Batty, M *et al.* (1994), Modeling inside GIS. Part 1: Model structures, exploratory spatial data analysis and aggregation, *International Journal of GIS*, vol. 8, pp. 291-307

Bernus, Peter & Laszlo Nemes (1996), *Modelling and methodologies for enterprise integration*. In Proceedings of the International Federation for Information Processing (IFIP) Conference on Models and Methodologies for Enterprise Integration, Chapman & Hall.

Berry, J.K. (1987), Fundamental operations in computer-assisted map analysis, International Journal of Geographical Information Systems, vol. 1, no. 2, pp. 119-136.

Bhat, M.V. (2007), Managing land information, *Coordinates*, vol. III, issue 8, pp. 22-7.

Bhat, M.V. & Bal Krishna (2002), *Survey of India – Towards a Contemporary Renaissance*, Centre for Spatial Database Management and Solutions and Department of Science & Technology, New Delhi (mimio).

Bishr, Y., W. Kuhn, and M. Radwan (1999), *Interoperating Geographic Information Systems, chapter Probing the semantic content of information communities: a first step toward semantic interoperability*, Kluwer Academic Publishers, pp. 211-21.

Blakemore, Michael (2004), Globalisation, infrastructure and agenda, *GIM International*, vol.18, No. 1, pp. 11-3.

Brich, T.W. (1949), *Maps: Topographical and Statistical*, Clarendon Press, Oxford.

Bricker, Charles (1968), *A History of Cartography, 2500 Years of Maps and Mapmakers*, Thames & Hudson, London.

Brotton, Jerry (2012), *A History of the World in 12 Maps*, Viking, New York.

Brotton, Jerry (2012), *A History of the World in 12 Maps*, Viking, New York.

Brown, Lester R. (1972), *World Without Borders*, Affiliated East-West Press Pvt. Ltd., New Delhi.

Buehler, Kurt (ed) (2003), OpenGIS Reference Model. Version: 0.1.2, No. OGC 03-040, Open GIS Consortium Inc.

Burrough P.A. & McDonnell, R.A. (1998), *Principles of geographical information systems: Spatial Information Systems and Geostatistics*. Oxford University Press.

Chatterjee, S.P. (1947), *The Partition of Bengal: A Geographical Study*, Calcutta Geographical Society, Publication No. 8, Calcutta.

Chaudhuri, S.B. (1964), *History of the Gazetteers of India*, Ministry of Education, Government of India, New Delhi.

Chawla, Rajiv & Subash Bhatnagar (2003), Bhoomi: A boom for farmers, *Geospatial Today*, vol. 1, issue 5, pp. 13-7.

Clarke, D. (2000), The global SDI and emerging nations - challenges and opportunities for global cooperation, *15th UNRCC Conference and 6th PCGIAP meeting*, 11-14 April, Kuala Lumpur.

Coleman, D.J. & J. McLaughlin (1998), Defining global geospatial data infrastructure (GGDI) : components, stakeholders and interfaces, *Geomatics Journal*, Canadian Institute of Geomatics, vol. 52, No. 2, pp. 129-44.

Corbett, J.P. (1979), *Topological Principles in Cartography*, Technical Paper No. 48, US Bureau of Census, US Government Printing Office, Washington.

Dasgupta, A.R. (2003), Spatial data infrastructure: Concepts and efforts, *Geospatial Today*, vol.2, issue 4, pp. 16-20.

East, W. Gordon & O.H.K. Spate (1953), *The Changing Map of Asia: A Political Geography*, Methuen & Co., London; and E.P. Dutton & Co., New York.

F.G.D.C. (1997), *Framework, Introduction and Guide*, Federal Geographic Data Committee, Washington, p. 106. Frank, S.M., M.F. Goodchild, H.J. Onsrud & J.K. Pinti (1996), *Framework Datasets for the NSDI*, The Federal Data Committee, Washington, D.C.

Feedback Strategies (2004), *Study to Assess the Size and Potential of the Market for Geospatial Data and Services: Survey of India,* Department of Science & Technology, Final Presentation (Mimio.).

Forghani, Alan & Denise Gaughwin (2002), Spatial information technologies to aid archaeological site mapping, *GIS@Development,* vol. VI, issue 6, pp. 31-5.

G.S.D.I. (1997), *Global Spatial Data Infrastructure: Conference Finding and Resolutions,* Chapel Hill, North Carolina, 21ˢᵗ October.

Gaba, I.D. (2002), *Integrated Personnel Policies: Survey of India,* Department of Science & Technology, New Delhi.

Garg, P.K. (2001), The changing paradigm of governance, *GIS Development,* vol. 5, issue 5, pp. 21-3.

Gavin, Elizabeth (2003), Roadmap to SDI in Africa, *GIS@Development,* vol.7, issue 9, pp. 29-31.

Georgiadeou, Yola (2003), Back to future, *Geospatial Today,* vol. 2, issue 4, pp.21-3.

Georgiadou, Yola (2003), Is Survey of India ready for the NSDI? *GIS@Development,* vol. 7, pp. 38-9.

Geospatial World (2022), *Geospatial Knowledge Infrastructure: Readiness Index and Value Proposition in World Economy, Society, and Environment,* Published jointly with United Nations Statistical Division.

GIS Development (2000), Internet GIS software, *GIS Development,* vol. IV, issue 7, pp. 28-32.

GIS Development (2000), Mapping milestones, GIS Development, vol. IV, issue 1, pp. 18-35.

Geospatial Today (2007), Facilitating spatial data, NSDI way, vol. 6, issue 5, pp. 26-7.

Geospatial Today (2007), NATMO – Born to map, vol. 5, issue 5, pp. 16-8.

Groot, Richard (2000), Corporatisation of national mapping agencies: challenges and opportunity, paper presented to the 15ᵗʰ UN Regional Cartographic Conference for Asia and the Pacific, Kuala Lumpur, 10-14 April.

Groot, Richard & John McLaughlin (2000), *Geospatial Data Infrastructures: Concepts, Cases and Good Practices,* Oxford University Press, New York, 1ˢᵗ ed.

Gulatee, B.L. (1954), The Height of Mount Everest, Technical Paper No. 8, Survey of India, Dehra Dun. Reprinted in 2020.

Gupta, Ravi (2000), GIS in the internet ear: What will India gain? *GIS Development*, vol. IV, issue 7, pp. 37-40.

Hadley, Clare & John Hammond (2003), Survey of India – A view from abroad, *GIS@Development*, vol. 7, pp. 29-31.

Hall, G. Brent (2005), Curriculum development in university level GIS education, *GIS Development*, Middle East edition, vol. 1, issue 1, pp. 18-21.

Huxhold, William E., Eric M. Fowler &Brian Parr (2004), *ArcGIS and the Digital City*, ESRI Press, Redlands, California.

Imhof, E. (1951), *Terrain et Carte*, Zurich.

Interview with Dr Prithvish Nag, Surveyor General of India for *Geospatial Today*.

ISO/IEC (1996), *Information Technology - Open Distributed Processing - Reference Model: Foundations. Draft International Standard,* International Organization for Standardization, Geneve, Switzerland, No. 10746-2.

James, Preston E. & Hibberd V.B. Kline, Jr. (1959), *A Geography of Man*, Ginn & Company, Boston, New York, London (Appendix A: Maps, pp. 489-520).

James, T.H. (1996), *The Theory of the Photographic Process*, MacMillan Co., New York, 3rd ed.

Jennings, Ken (2011), *Maphead: Charting the Wide, Weird World of Geography Wonks*, Scribner, New York.

Jensen, J.R. (1996), *Introductory Digital Image Processing: A Remote Sensing Perspective*, Prentice Hall, New Jersey, 2nd ed.

Kaplan, Robert D. (2012), *The Revenge of Geography*, Randon House, New York.

Kasturirangan, K. (2009), Shaping the NSDI, in   Maj Gen (Dr) R. Siva Kumar (ed), *Indian NSDI: A Passionate Saga*, NSDI Directorate, New Delhi. pp. iv-vii.

Keay, John (2000), *The Great Arc*, Harper Collins, London.

Kelmelis, John A. (2002), The national map, its scientific evolution, *GIS India*, vol. 11, No. 12, pp. 6-12.

Kottman, C. (1999), *Interoperating geographic Information Systems*, (chapter on *The OpenGIS Consortium and progress Toward Interoperability in GIS*). Kluwer Academic Publishers.

Kozakis, John T. (1959), World topographic map coverage, in the *Proceedings of IGU Regional Conference in Japan 1957*, The Organising Committee of IGU Regional Conference in Japan, The Science Council of Japan, Tokyo, pp. 553-7.

Krishna, Bal (2003), Breaking inertia – Survey of India embraces a new era, *GIS@Development*, vol. 7, pp. 40-2.

Kumar, G.S. & Monika Patwa (2002), WWW.surveyofindia.gov.in, *GIS India*, vol. 11, No. 12, pp. 15-6.

Lawrence, Vanessa (2002), Mapping out a digital future for Ordnance Survey, *The Cartographic Journal*, vol. 39, No. 1, pp.77-80.

Lemmens, Mathias (2003), Free of charge or revenue generation, Invited Reply, *GIM International*, February, pp. 40-9.

Lieberman, Josua (ed) (2003), OpenGIS Web Services Architecture. Discussion paper, Version: 0.3, No. OGC 03-025, Open GIS Consortium Inc.

Lorena, Montoya & Ian Masser (2003), GIS for urban disaster management: facing the unexpected, *Geospatial Today*, vol.2, issue ,2 pp. 12-5.

Luisa Liliana, Alvarez C. (2003). *Geoinformation virtual enterprises design and process management*. Master's thesis, International Institute for Geo-Information Science and Earth Observation ITC, Enschede, The Netherlands.

Maclver, Robert (1964), *The Modern State*, Oxford University Press.

Maitra, Julie Binder (2001), The national spatial data infrastructure in the United States: standards, metadata, clearing house and data access, *Proceedings of the National Geospatial Data Infrastructure (NGDI): Towards a Road Map for India*, CSDMS, 5-6 February, New Delhi, pp. 6-11.

Maor, Eli (1999), *Trigonometric Delights*, Princeton University Press, Princeton.

Mapping Science Committee (1997), *The Future of Spatial Data and Society*, National Academy Press, Washington.

Markhan, Clements Robert (Sir)(1871), *A Memoir on the Indian Surveys*, W.H. Allen & Company, 2nd edition in 1878.

Markhan, Clements Robert (Sir)(1895), *Major James Rennel and the Rise of Modern English Geography*, Cassell & Company.

Masser, I. (1998), *Governments and Geographic Information*, Taylor & Francis, London.

McGrath, Gerald (1982), Redefining the role of government in surveys and mapping: A view of events in the United Kingdom, *Cartographia*, vol. 19, Nos. 3&4, pp. 44-52.

Ministry of External Affairs (1960), *Atlas of the Northern Frontiers of India*, Government of India, New Delhi.

Ministry of External Affairs (u.d.), *India & Bangladesh: Land Boundary Agreement*, Public Diplomacy Division, Government of India, New Delhi.

Mitasova, *et.al* (1995), Modeling spatially and temporally distributed phenomena: new methods and tools for GRASS GIS, *International Journal of GIS*, vol. 9, pp. 433-46.

Moellering, H., H.J.G.L. Aalders & A. Crane (eds) (2003), *World Spatial Metadata Standards,* International Cartographic Association.

Monkhouse, F.J. & H.R. Wilkinson (1952), *Maps and Diagrams: Their Compilation and Construction*, Methuen & Co., London and E.P. Dutton & Co., New York.

Mott, Peter (1990), Fifty years ago-an account of the 1939 expedition to the western Karakoram, in the Souvenir published on the occasion *Everest Bi-centenary Celebration*, Survey of India, Dehra Dun, 4-5 October 1990 (mimio).

Morales, Javier & Mostafa Radwan (2002), Extending geoinformation services: A virtual architecture for spatial data infrastructures, in *Proceedings of the Joint International Symposium on GeoSpatial Theory, Processing and Applications,* ISPRS Commission IV, Ottawa, Canada.

N.S.D.I. (1994), *Coordinating Geographic Data Acquisition and Access*, The National Spatial Data Infrastructure. Executive Order No. 12906, Executive Office of the President of the US.

Nag, P. & Smita Sengupta (2007), *Geographical Information System: Concepts and Business Opportunities*, Concept, New Delhi.

Nag, P. (1986), A proposed base for geographical information system for India, *International Journal for Geographical Information System,* Taylor & Francis, London, vol. 1, No. 2, pp. 181-7.

Nag, P. (1997), *Geomorphological Mapping*, National Atlas & Thematic Mapping Organisation, Kolkata.

Nag, P. (1999), Getting smarter with Smarts maps, *GIS Development*, January-February, pp. 20-4.

Nag, P. (2002), Maponomics – Map products and their commercialisation,

GIS@Development, vol. 6, issue 2, pp. 19-23.

Nag, P. & B. Nagrajan (2003), From GTS to GPS. *GIS@Development*, vol. 7, issue 12, pp. 15-6.

Nag, P. (2007), Geomatics and GIS: Definition and scope, *Lecture Notes of Training Course on Spatial Data Management for PURA Related Development Initiatives*, SPCMF and NATMO, Kolkata.

Nag, P. (2008), Transformation of national mapping agencies: The case of India, *Indian Journal of Landscape Systems and Ecological Studies (ILEE)*, vol. 31 (1), pp. 1-6.

Nag, P. (2014), *Vision Document: Natural Resource Data Management System*, Department of science & Technology, New Delhi.

Nag, P. (2016), *Indian Geospatial Infrastructure*, Bharati Prakashan, Varanasi.

Nag, P. (ed) (1984), *Census Mapping Survey*, IGU Commission on Population Mapping, Concept Publishing Company, New Delhi.

Nag, P. (ed) (1992), *Thematic Cartography and Remote Sensing*, Concept Publishing Company, New Delhi.

Nag, P. & B. Nagrajan (2003), Geodesy and gravimetry, *Indian National Report for IUGG 2003*, Indian National Science Academy, New Delhi, pp. 5-17. Submitted to the Twenty Third General Assembly of the International Union of Geodesy & Geophysics, Sapporo (June 30 – July 11, 2003).

Nag, P. & G.C. Debnath (2021), *An Advanced Geography of India*, Bharati Prakashan, Varanasi.

Nag, P. & H.B. Madhwal (2006), Geodetic and geophysical studies of crustal deformation and fault displacements- A case study for monitoring deformation in Kachchh region of Gujarat, *Geographical Review of India*, vol. 68, No. 2, pp.174-85.

Nag, P. & M. Kudrat (1998), *Digital Remote Sensing*, Concept Publishing Company, New Delhi, pp. 129-40.

Nag, P. & Smita Sengupta (2007), *Introduction to Geographical Information System*, Concept Publishing Company, New Delhi.

Nag, P. & Smita Sengupta (2008), *Geographical Information System: Concept & Business Opportunities*, Concept Publishing Company, New Delhi.

Narayan, L.R.A. (u.d.), *Surveying and Mapping the Great Himalayas – A Surveyor's Challenge*.

National Geographic (2008), How India moved: Plate tectonics and India's landscape, *The Geography Teacher*, vol. 5, No. 1, pp. 18-21.

NIMA (2000), *World Geodetic System 1984*, Technical Report TR 8350.2, 3rd ed., Amendment 1, National Imagery & Mapping Agency, Department of Defence.

O.S.D.M. (2002), *Office of Spatial Data Management Glossary*, Online: http://www.Osdm.gov.au/osdm/glossary.html.

Oberlander, Theodore M. (1968) A critical appraisal of the inclined contour technique of surface representation, *Annals of the Association of American Geographers*, Vol. 58, No. 4, pp. 802-13.

Ordnance Survey (1999), *Annual Report and Accounts: 1988-99*, The Stationary Office, London.

Ordnance Survey (2001), *Quinquennial Review of Ordnance Survey*, Stage 1 Report by CMG Admiral, Surrey.

Ordnance Survey (2002), *Annual Report and Accounts: 2001-2002*, The Stationary Office, London.

Ormeling, F.J.O. (1982), ICA Report-1980-82, *Cartographia*, vol. 19, Nos.3&4, pp.102-104.

Owen, Tim & Elaine Pilbeam (1992), *Ordnance Survey: Map Makers to Britain Since 1791*, Ordnance Survey, Southampton.

P. Nag (2005), Water and land in Andaman and Nicobar Islands, *Indian Journal of Landscape Systems and Ecological Studies (ILEE)*, vol. 28 (2), pp. 13-20.

Pande, Amitabha (2003), Market needs to play a key role in influencing policy, *GIS@Development*, vol. 7, pp. 46-7.

Pande, Amitabha, R. Siva Kumar, Akhsay Jaitley & Saketh Shukla (2003), NSDI – The legal prescription, *Geospatial Today*, vol. 2, issue 4, pp. 24-32.

Panikkar, K.M. (1955), *Geographical Factors in Indian History*, Bharatiya Vidya Bhavan, Bombay.

Parson, E.D. (2006), Embedding geospatial technology into mainstream, *GIS Development*, vol. 10, issue 2, p. 18-20.

Percivall, George (ed)(2002), The openGIS Abstract Specification Topic 12: Open GIS Service Architecture. Version 4.3, Open GIS Consortium, Inc.

Phillimore, R.H. (1946-58), *Historical Records of the Survey of India*, Survey of India, Dehra Dun (Volume I: up to 1800; Volume II: 1800-15; Volume III: 1815-30; Volume IV: 1830-43; and Volume V: 1843-45).

Prabhu, Suresh & Shobhit Mathur (2023), We need biographies written of our great academic institutions, *Mint*, 5th January.

Prabhu, Suresh & Shobhit Mathur (2023), We need biographies written of our great academic institutions, *Mint*, 5th January, p.12.

Prasad, Manish (2001), Product development, *GIS Development*, vol. 6(8), p. 19.

Publication Division (1964), *The Chinese Threat*, Ministry of Information & Broadcasting, Government of India, New Delhi.

Pulsani, Preetha (2001), Internet GIS – One perspective, *Map India 2001 Conference Proceedings*, 7-9 February, pp. 19-22. Summary.

Pulusani, Preetha (2003), Interoperability – Trend or reality? *GIS@Development*, vol. 7, issue 9, pp. 16-9.

Raisz, Erwin (1948), *General Cartography*, McGraw-Hill Book Company, New York, Toronto, London.

Rajabifard, A. & I. P. Williamson (2002), Spatial Data Infrastructures: An initiative to facilitate spatial data sharing, in R. Tateishi & D. Hasting (eds), *Global Environmental Databases – Present Situation and Future Directions*, vol. 2, International Society for Photogrammetry and Remote Sensing (ISPRS-WG IV/8), GeoCarto International Centre, Hong Kong.

Rajabifard, A. Feeny & I.P. Williamson (2002), Future directions for SDI development, *International Journal of Applied Earth Observation and Geoinformation*, vol. 4, No. 1, pp. 11-22.

Rajabifard, A., F. Escobar & I.P. Williamson (2000), Hierarchical spatial reasoning applied to spatial data infrastructures, *Cartography Journal*, vol. 29, No. 2, Australia.

Rajabifard, A., I.P. Williamson, I.P. Holland & G. Johnstone (2000), From local to global SDI initiatives: A pyramid building blocks, *Proceedings of the 4th Global Spatial Data Infrastructures Conference*, 13-15 March 2000, Cape Town, South Africa.

Rajabifard, A., T.O. Chan & I.P. Williamson (1999), The nature of regional spatial data infrastructure, *Proceedings of the AURISA 99*, pp. 22-6, November 1999, Blue Mountains, NSW, Australia.

Ramachandran, R. (2003), Celebrating a survey saga, *Frontline*, July 4, pp. 107-12.

Ramamurthy, K. (1982), *Map Interpretation (Indian Landscapes through Survey of India Topographic Maps)*, Madras.

Rao, M. & K.R. Sridhara Murthi (2012), *Perspectives for A National GI Policy;* *National Institute of Advanced Studies,* IISc Campus, Bangalore September, p.124.

Rao, Mukund & V. Jayaraman (1995), *Guidelines for GIS Standardisation,* ISRO-NNRMS-TR-105-95, Indian Space Research Organisation, Bangalore.

Remkes, J.W. (2000), Foreword in R. Groot & J. McLaughlin (eds), *Geospatial Data Infrastructure - Cases, Concepts and Good Practices,* Oxford University Press, New York. Rhind, D. (1999), *Key Economic Characteristics of Information,* Ordnance Survey, United Kingdom.

Rhind, D. (ed) (1997), *Framework for the World,* Geoinformation International, Cambridge.

Rhind, David (1981), Geographical information system in Britain, in N. Wringley & R.J. Bennet (eds), *Quantitative Geography: A British View,* Routledge & Kegan Paul, London.

Rhind, David (1989), Why GIS? *ARC News,* vol. 11(3), ESRI, Redlands, California.

Rhind, David (2000), Facing the challenges: Redesigning and rebuilding Ordnance Survey, in David Rhind et al (ed), *Re-inventing Government & NMOs,* London, 275-304.

Rhind, David (2001), Lessons learnt from local, national and spatial data infrastructure, *Proceedings of the National Geospatial Data Infrastructure (NGDI): Towards a Road Map for India,* CSDMS, 5-6 February, New Delhi, pp. 26-30.

Rhind, David (2005), GIS education for a changed world, *GIS Development,* Middle East edition, vol. 1, issue1, pp. 14-7.

Robinson, Arthur H. (1954), Geographical cartography, in Preston E. James & Clarence F. Jones (eds), *American Geography: Inventory & Prospects,* Association of American Geographers, Syracuse University Press, pp. 553-77.

Robinson, Arthur H. (1960), *Elements of Cartography,* 2nd Ed., New York.

Roy, P.S. & Samir Saran (2003), An interoperability model for Indian NSDI, *Geospatial Today,* vol. 2, issue 4, pp. 36-8.

S.D.I. (2000), *Developing Spatial Data Infrastructure: The SDI Cookbook,* Version 1.0, Online: http://www.Gsdi.org/pubs/cookbook/cookbook0515.pdf.

Saraf, Madhav N. (2005), GIS-GPS academics in India: Present and future, *GIS Development*, vol. 9, No.1, p p.. 32-3.

Sen, Sumit (2007), Interoperability – key to sustaining NSDI, *Geospatial Today*, vol. 6, issue5, pp. 28-32.

Shabad, Theodore (1956), *China's Changing Map*, Methuen & Co., London.

Siddartha, K. (2013), *Geography through Maps*, Kisalaya Publication, New Delhi.

Sipkes, Jacques (2003), The Bhoomi project: computerisation of land records, *GIM International*, vol 17, No, 6, pp. 51-3.

Sipkes, Jacques (2004), AfricaGIS 2003: How far have we come? GIM International, vol.18, No. 1, pp.55-9.

Skelton, R.A. (1958), *Explorers' Maps*, Routledge & Kegan Paul, London.

Srivastava, Rudra Prakash (1996), The changing map of India, in L.R. Singh (ed), *New Frontiers in Indian Geography*, Professor R.N. Dube Foundation, Allahabad (Prayagraj), pp. 114-29.

Survey of India (1920-21), *The Imperial Atlas of India, Political Edition*, published by the Surveyor General of India, Calcutta.

Survey of India (1982), *Records of Technical Committee Meeting*, Departmental Paper 20, Dehra Dun.

Survey of India (2002), *Towards a Contemporary Renaissance, Background paper for discussion on Restructuring of Survey of India* (mimio).

Swain, P.H. & S.M. Davis (eds) (1978), *Remote Sensing: The Quantitative Approach*, McGraw Hills, London & New York.

Task Force on NSDI (2002), *National Spatial Data Infrastructure (NSDI): Strategy and Actional Plan*, Department of Science & Technology, New Delhi.

Thomas, George B. & Finney, Ross (1979), *Calculus and Analytical Geometry*, 5th ed., Wesley. Tomlinson, R.F., H.W. Calkins & D.F. Marble (1976), *CGIS: A Mature, Large Geographic Information System*, UNESCO Press, Paris.

Tooley, Ronald Vere (1978), *Maps and Map-makers*, Batsford, London 6th edition.

Unknown (1826), *Index Containing the Names and Geographical Positions in the Maps of India*, Kingsbury, Parbury & Allen, London.

U.S.G.S. (2005), *Geographic Information System*, United States Geological Survey, Reston.

Venketachalam, P.B., P.B. Krishna Mohan, J.K. Suri, Aarthi T. Chandrasekar & Vikas Mishra (2001), Teaching GIS principles through multimedia based GIS tutor, *GIS Development*, vol 5, No. 1, pp. 24-7.

Vijyan, Suchitra (2021), *Midnight's Border: A People's History of Modern India*, Westland Publication, Chennai.

Wade, Tasha & Shelly Sommer (2006), *A to Z GIS: An Illustrated Dictionary of Geographic Information Systems*, ESRI Press, Redlands, California.

Waters, N. (2001), Internet GIS: Watch your ASP, *Geoworld*, vol. 14, No. 6, pp. 26-8.

Williamson, Ian, Abbas Rajabifard & Mary-Ellen F. Feeney (eds)(2003), *Developing Spatial Data Infrastructures : From Concept to Reality*, Taylor & Francis, London and New York.

Wood, Denis (1992), *The Power of Maps*, Guilford Press, New York.

Wu, Jian Kang (2006), Embedded GIS – An overview, *GIS Development*, vol. 10, issue 2, pp. 28-30.

Zeiler, H. (1999), *Modeling Our World: The ESRI Guide to Geodatabase Design*, ESRI, Redlands, California.

Quotations & Comments

1. **Clifford J. Mugnier, The public trust of surveys & maps, Grids and Datums,** *Photogrammetric Engineering & Remote Sensing,* **November 2001, pp. 1229-30.**

    "The Survey of India has operated for most of its history as a military intelligence operation. In the early years, agents were dressed in disguises at times in order to garner approximate locations of villages and towns, information on local roads and topography, etc. The need for secrecy at the time was quite real, because the British were trying to wrest political control from the many local maharajas. That basic fabric of military secrecy still pervades the Survey of India to this day. The Director is a flag-rank officer in the Army. Recurring border skirmishes over contested territory add to the military's perceived need for secrecy in India. The availability of surveying and mapping data is quite problematic nowadays for peaceful use by Indian citizens. There's simple rule: "you may not have the data.""

2. **Kapil Raj, Mapping knowledge go-betweens in Calcutta, 1770-1820,** *The Brokered World,* **in Chapter 3, pp. 133-4.**

    "It was also much more: in order to compile his 'Map of Hindoostan', material for which was garnered within the institutional context of the Surveyor's office and later in London, he enjoyed not only the close collaboration of Boughton Rouse, but also benefited from his relations with a fellow army officer, Robert Barker (c.1732–1789) who, in the course of his military campaigns in Bengal and the Philippines and sundry private trade enterprises, had acquired a sizeable fortune and skills of mediation which gave him privileged access to various knowledge milieux ranging from Benares astronomer-pandits to the increasingly dispersed Mughal Persianate literati, and eventually to the Royal Society (to which he was elected in 1775, on his return to England). Others helped procure relevant information for him from other parts of the subcontinent, such as the map of Gujarat made by 'a Brahmin of uncommon genius named Sadānand' or the route map from Bengal to the Deccan made by a sepoy, Ghulām Muhammad. The intermediaries also included Frenchmen turned 'Turk', such as Antoine Polier and Claude Martin. In addition, identifying contemporary place names (with their pronunciations varying between different groups of speakers) and establishing a correspondence between their appellations with those known in Greek and Ptolemaic geography meant dabbling in philology, which he got from William Davy (d. 1784) who, as an ensign in the Company's army, had learnt Persian and whose latinized system of orthography was subsequently adopted not only by the India Surveys, but also in faraway Oxford, for English transcriptions of Persian. Jonathan Scott, who during his years as an ensign in the ELC's army in the 1700s had also learnt Persian and Hindustani and on his return to England translated and published the memoirs of a late Mughal nobleman in 1786, provided Renell with an invaluable source for writing the political geography of Hindustan for the 1788 edition of the *Memoir of a map of Hindoostan*. Without their mediation, it is hard to imagine Renell gaining access to any of the knowledge circuits so crucial to the compilation of his innovative map."

(References quoted)

- James Renell, *A Bengal atlas: containing maps of the theatre of war and commerce on that side of Hindoostan* (London, 1780).

- James Renell, *Memoir of a map of Hindoostan, or the Mogul empire* 3rd ed. (London, 1793), p. 185, n. 6. For Renell in Bengal, see Thomas Henry Digges la Touche, ed., The journals of Major James Renell, first Surveyor-General of India, *Memoirs of the Asiatic Society of Bengal* 3 (1910), pp. 95–248; Andrew S. Cook, 'Major James Rennell and *A Bengal atlas* (1780 and 1781),' *India Office Library and Records, Report 1976* (London: HMSO, 1978), pp. 4–42.)].

3. **R. Ravindranath (2004), Indian Classics – 150 Years, *The Hindu*, October 9, 2004.**

"The Governor General Dalhousie put pressure on Capt. Thuillier, the Deputy Surveyor General in charge of the lithographic departments of the Surveyor General Office in Calcutta. H.M. Smith prepared new designs for the stamps with the head of Queen Victoria in four values, i.e. ½ anna, one anna, four annas and eight annas."

"Capt. Thuillier and engraver Muniroodin began to work on the first bi-colour stamp in the world. Greater care was needed, as it was a two-colour job."

"What started in 1854 as a trail is now deeply rooted in Indian philately and 150-year saga of Indian postage stamps is a fascinating one."

4. **Robert D. Kaplan (2012), *The Revenge of Geography: What the Maps Tells us about Coming Conflicts and the Battles Against Fate*, Random House, New York, p. 233.**

"It is not that human settlement from early antiquity forward doesn't adhere to subcontinental geography; rather, it is that India's geography is itself subtle, particularly in the northwest, telling a different story than the map reveals at first glance, the relief map shows a brown layer of mountains and tableland nearly marking off the cool wastes of middle Asia from the green tropical floor of the subcontinent along the present border between Afghanistan and Pakistan."

5. **Ken Jennings (2011),** *Maphead: Charting the Wide, Weird World of Geography Wonks*, **Scribner, New York, p. 88.**

    "Now please try to resist blowing yours brains out when I tell you that the Great Trigonometrical Survey that mapped British India two centuries ago required more that forty thousand triangles to complete and stretched from a five-year project into an eighty-year one."

    Footnote: The first survey of this kind was begun in France in the 1670s by Giovanni Cassini, and it proved so daunting that his grandson wounded up finishing it more than a century later. This was the first topographical map of an entire nation ever made, but it revealed France to be much smaller in area than it had always been drawn. "Your work has cost me a large part of my state!" King Louis XIV reportedly huffed.

6. **Keay, John (2000),** *The Great Arc*, **HarperCollins Publishers, London, p. xvii.**

    "That the world's highest point is in fact called after George Everest, a controversial British Colonel who had never even seen the mountain, let alone climbed it, first dawned on me when I was writing a book about the exploration of Kashmir. Everest did not feature in the region, either as man or mountain, but as an institution, dear to the Colonel's heart and known as the Survey of India, did. Most of Kashmir, including the Karakoram mountains, had first been measured and mapped by the men of the Indian Survey. And the Survey being a government department within British India's bureaucratic Leviathan, it had generated copious records."

7. **India's "Great Arc" lights up Cambridge,** *Newindress.com*, **July 23, 2003:**

    "CAMBRIDGE (ENGLAND): Mapping experts from 71 countries have gathered here for their four-yearly conference, traditionally hosted by Britain's Ordnance Survey.

    Their theme is "National Mapping Shaping the Future" but the highlight is an exhibition that celebrates an achievement of the past- the geographical survey that began in Madras (now Chennai) 200 years ago and reached its goal in the Himalayas 40 years later.

    The exhibition of "The Great Arc" is touring Britain for six months. On Sunday, it marked the opening of the international mapping conference, and the day was hailed as "India Day" in honour of the most remarkable achievement of the Survey of India – the Herculean Task of surveying the sub-continent of "Hindoostan" along the 78th meridian

in accordance with the most precise and correct mathematical principles."

8.   Muthiah, S. Nor Recalled Here still, *The Hindu*, 30ᵗʰ June 2003:

"Meanwhile, I know of at least one person, T.S. Subramaniam, who has
been scrambling up hill and down to find markers of that survey in
Madras. St. Thomas' Mount has provided a likely marker that is being
speculated over, as inscription on it there's none. Certainly, the base
of the marker of Lambton's starting point, near the church in a south-
westerly direction, was in place in 1885/6 but had vanished by 1915,
according to a Survey of India document of the time. A GTS station
mentioned in the same report as being "on the southwest wing of the
Parry Building' could not have been in *Dare House* (built 1939-41) and
may not have even been in the old Parry Building, which was developed after a Nawab of Arcot property here was bought by Parry's in
1803 and later developed. And GTS stations mentioned at Red Hills
and Injambakkam have disappeared. All that remains as a memorial to
Lambton's work is a 15-foot tall granite pillar in the Regional Meteorological Centre, Nungambakkam, which is not a GTS station, and a GTS
benchmark a few metres away.

The inscription on the pillar, in English, Tamil, Telegu, Urdu and Latin
reads:" (1) The Geodetic position (Lat. 13° 4'3" 0.5 N, Long. 80° 14'
45" 20 E) of Col. William Lambton, primary original of Survey of India
fixed by him in 1802, was at a point 6 feet to the south and 1 foot to the
west of the centre of this pillar. (2) The centre of the meridian circle of
the Madras Observatory was at a point 12 feet to the east of the centre
of this pillar." This inscription is believed to have been cut after astronomer Michie-Smith had made the final longitudinal determination on
Madras in 1892."

9.   Gajendran, K., A Boost to Survey and Mapping: Interview with Prof V.S. Ramamurthy, *Frontline*, vol. 20, Issue 13, June 21 – July 04, 2003, pp. 1-5:

*"How is the survey conducted 200 years ago relevant now, for there have been
so many advances in mapping tools and techniques?*

The relevance comes from the fact that the scientific basis of this entire experiment has not changed while technology has changed. You
see this several areas. Our entire communication system today is electro-magnetic and digital, but the technology has changed. The digital
revolution happened 20 years ago and is getting upgraded. The basic
principles of mapping and surveying have set at that time and new
technologies are coming into the picture. It is important because there
is greater realisation that the domain in which maps will be useful has

considerably broadened. Therefore we have to tell our people that we have a tradition in mapping so that everybody becomes familiar with the kind of knowledge maps provide them."

10. Dhaundiyal, Pallavee, *Mapping India to glory*, Times News Network, Times of India, 22nd April 2002:

Almost 200 years ago, Col William Lambton planned his itinerary to walk the nerve centre of the steaming Indian land. On April 10, 1802, he finally began the longest measurement of the earth's surface- the length of a degree of latitude along a longitude in the middle of penin-sular India, at St Thomas Mount in Madras and marked the beginning of great trigonometrical survey.

The ensuing 50 years witnessed inch-perfect survey of over 2,400 kms. The Great Indian Arc, that grew over 50 years, made possible the map-ping of entire Indian subcontinent and consequently, its infrastructural development including railways, national highways, telephone lines and power grids. It attempted to compute the precise curvature of the globe and significantly advanced our knowledge of the exact shape of our planet. The Arc also resulted in the first accurate measurement of the Himalayas that was acknowledged by the naming of the world's highest peak in honour of Col Sir George Everest.

April 10, 2002, marked the bicentennial of the described scientific un-dertaking. The Department of Science and Technology, and Survey of India have announced a year long celebrations and educational events both in India and UK. Besides, the Surveyor General of India P Nag de-clared a comprehensive programme to revitalise, modernise and re-en-gineer the oldest institution of the Indian Government – The Survey of India.

Nag said, "The objective is to use the celebrations as an opportunity to leverage the wealth of data held by the Survey of India and convert them into knowledge products to meet the rapidly growing needs of the society." While highlighting the significance of the survey in the contemporary world, VS Ramamurthy, secretary of Department of Science and Technology said, "The great Trigonometric Survey has a contribution to all the topographical surveys ever conducted in the country along with making huge impact on the development of science and technology of today.

Announcing the major events to be held during the year, Ramamur-thy further said: "In order to enthral and infuse interest in the field of spatial sciences, the events for the year include the treasure quest, geo quest quiz, Great Arc exhibition, Great Arc documentary film series, Great Arc pictorial publication and many more."

The series of events first commenced in the capital city with the treasure quest for students. Just as Sir William Lambton and George Everest set out on their quest for mapping the entire Indian subcontinent exactly 200 years ago, 16 budding school students from Delhi set out on a similar expedition, in their quest to map seven cities that constitute the city of Delhi.

The event offered the students an opportunity to challenge the limits of their historical knowledge and skills at using maps. The 'quest' featured four teams from the Modern School, Barakhamba Road; the Tagore International School, East of Kailash; the Ludlow Castle and the Kendriya Vidyalaya, RK Puram. Each team was provided a set of clues that marked historic landmarks of the city. The 'hunt' took the teams across the city and witnessed an intense and frantic, yet competitive spirit among all participants. Ultimately, the triumphant team was of the Tagore International School, which reached the treasure chest at the Qutab Minar, moments before their nearest rivals Ludlow Castle. They were awarded gold coins by Murli Manohar Joshi, Union Minister of Human Resource development, Science and Technology.

11. **Editorial, Mapping Defence, *The Telegraph*, 6th May 2003, vol XXI, No 299, p. 12:**

There are still people around who believe that the earth is flat. If this sounds incredible one has only to look at India's ministry of defence. The ministry retains control over the production of maps. In the latest instalment of its show of control, the defence ministry is likely, according to reports, to shoot down the idea of producing computer-based navigational maps which could help motorists and travellers to find their bearings in an unknown area or locality. According to existing regulations the production of all maps, in print or in digital form, require the prior approval of the ministry of defence and the Survey of India. This approval is not always forthcoming and publishers of maps have to go through innumerable bureaucratic obstacles to get the required permission. The project that is now under the threat of being vetoed by defence ministry officials is similar to the global positioning system which is now widely used all over the developed world. It will be recalled that earlier this year, the defence ministry, in a fit of paranoia, ordered the Survey of India to do away with contour lines and the highlighting of intersection points between latitudes and longitudes. These essential features of a map are supposed to represent, according to the wisdom of defence ministry mandarins, a threat to national security since missile technology uses contour lines.

Maps have always been very prickly things for Indian officialdom. The rejection of the GPS is one more instance of the bizarre and antiquated notions that inform the decisions of people who nonchalantly walk the corridors of power. Any power or organization harbouring hostile intentions towards India will not be insane enough to depend on contour lines on maps or on the GPS to execute their plans. Far more sophisticated technology and methods are available to get data about the enemy's terrain. One hopes for the sake of India's security that the ministry of defence and intelligence agencies use this technology and do not depend on contour lines and the GPS. The control that the defence ministry and the Survey of India maintains over maps and their production is nothing more than a show of petty power. It is one of the ways in which the state stands in the way of citizens accessing even ordinary and innocent information. Indian officialdom is unaware that in an era when even the most secret and classified information of the Pentagon or the Central Intelligence Agency can be hacked into, this kind of obsession about withholding information is nothing short of obtuseness.

12. Molennar, Martien, Surveyors Need to Change their Attitude, *GIM International*, October 2001, p. 7:

"Many countries have mapping organisations within which the production processes for topographical mapping date from colonial times. Such organisations have a hard time to keep pace, in terms of production rates, with the fast development of their country. Traditional approaches are too slow. I think that along with new mapping techniques we should develop new concepts for core data production and provision which avoid the slow processes of traditional map production. If we do not dwell too much on interpretation but instead correlate data geographically sufficiently correctly with GPS, then we can work on the provision of a good core dataset capable of being linked to other types of information, like vegetation and soil data. In training and conceptual development there should come a move away from traditional map concept still so often found in the definition of products."

13. Ramachandran, R., Survey Saga, *Frontline*, May 10, 2002, pp. 66-70:

"Alongside, an effort is being made to enable free public access to topographical data – analogue as well as digital – without the prevalent security and defence-related restrictions on access. For this a mammoth exercise has been undertaken to map the geographic control points on a different geodetic datum called WGS-84 instead of Everest Spheroid, using satellite-based Global Positioning System (GPS) surveying techniques. While some digital topographic maps of southern India on 1:250,000 scale on WGS-84 should become available in the next three to four months, the mapping of the WGS-84 control points is expected to take three or four years and digitised topographical maps of the entire country are expected to be available for public access in seven to eight years."

14. The Statesman, 100 Years Ago, Today September 6, 1902. Reproduced in *The Statesman*, September 6, 2002, Vol. CXXXXIII 211:

"Most people who are impelled, by the exigencies of their profession, or by their interest in Indian affairs, to glance through the General Report of the Operations of the Survey of India, which has just been published, will probably find themselves wishing that, in this instance at will events, the well-known orders of Lord Curzon's Government regarding the curtailment of official reports had been less rigidly observed. To any reader, not destitute of imagination, the Surveyor General's report for the year 1900-01 furnishes something like the framework of a romance, but the exploits of the Survey officers are condensed into the briefest paragraphic statements, or setforth in figures, which are as far as possible from being figures of speech. "The style we like is the humdrum", said the old Director of John Company to the Ciceronian candidate Colonel St G.C. Gore's Department shares that unpretending taste. ....If it be true that few people in India are able to realise the nature and extend of the work that is being done by the Survey Department, the fact is not difficult of explanation. Take, for example, a paragraph which occurs in the summary of Topographical Surveys. It refers to No. 3 party, under the charge of Mr E. Litchfield in Lower Burma. "The party suffered severely from malarial fever throughout the field season, which has been the most unhealthy experienced since its formation. In Pegu not one man escaped fever during the first 3 and half months, and for the remainder of the season all suffered more or less from its effects, and were unfit for hard work. In Sundoway things were no better. Four men died, and many returned to their homes completely broken in health. Of the Sub-Surveyors recessing at Bangalore, several are still ill and are undergoing treatment." This is but one out of a dozen entries indicating with the eloquent plainness of the dangers which confront the missionaries of science, who, in every corner of India, are engaged in compiling the record of the country's physical characteristics."

(Courtesy: Maj Gen Pranab Dutt, AVSM, VSM, Dehra Dun).

15. *A Brief History of Survey of India, 1767 – 1967*, Survey of India, Dehra Dun (*Mimio.*), pp. 2 and 18:

"Many developing countries have been handicapped because of lack of proper map cover – we are fortunate we had the data collected in the past and a nucleus on which to build. A measure of the dependence of planned development on proper surveys is the department's potential had to be employed on surveys for various developmental projects, till we were confronted with Chinese aggression. Suddenly the peaceful northern frontier which had been considered impregnable from time immemorial became a live issue. The department played its humble role in providing up to data and accurate map coverage by diverting its major efforts to this task."

"Up to this time there was no reliable atlas of India. It was Collin Mackenzie who initiated the idea of compilation of an Indian Atlas series on a sufficiently large scale. In this he had in mind the maps of the Southern peninsula compiled on quarter inch scale by Arrow Smith, the famous cartographer. It was in 1823 that thee Directors wrote "We are extremely desirous of forming …. a complete Indian Atlas upon a scale of 4 miles to an inch …… the best suited to general purposes, and which has been adopted by Arrow Smith in a recent publication …. This map would ….. form a useful basis for complete geographical delineation of India."

16. *The Atlas of India* (1829), **published by the East India Company,** *Scientific Intelligence*, **p.347:**

This noble work, of itself a splendid monument of the munificence of the East India Company, is upon a scale of four miles to an inch, and taken from actual surveys, which when completed, will form a map of India on one uniform plan.

The project was first conceived by Colonel McKenzie, and a large portion of those parts already published were surveyed under his superintendence. The surveys on the northern part of the peninsula have for their basis the triangulation of Colonel Lambton, who extended a set of principal and secondary triangles over the whole country.

The sheets are published as they are completed; some of them have blank spaces, to be filled up as the surveys proceed; nothing being allowed to go forth to the world which is not founded upon actual survey. The following are the sheets already published.

Sheet 47 contains the surveys of Captain Hodgson and Lieutenant Herbert in the mountainous country comprising the northern part of the province of Sirmur, and the principal part of Bissahir.

Sheet 48 contains the surveys of Captain Hodgson and Lieutenant Herbert in the Southern part of the province of Sirmur, part of Garhwal, and the Debra Dún. The flat country is from the surveys of Lieutenant White, Lieutenant Hodgson,

Captain Colvin, and Lieutenant Blake. Sheet 65 is principally the survey of Captain Hodgson and Lieutenant Herbert of the sources of the Gunges, and of Captain Hearsey and Mr. Moorcroft of the sources of the Indus and Setluj.

Sheet 66 is principally the survey of Captain Webb, of the province of Kamaun. Sheets 69 and 70, contain the greater part of the province of Bandelkund surveyed by Captain Franklin, brother of Captain Sir John Franklin, R. N.

Sheets 42, 43, 58, 59, 6O, 77, 78, 80, 81, 95, are surveys executed in the peninsula.

--As. Jour.

17. **Ramamurthy, V.S. (2004), The NSDI is transforming itself from paper to reality,   *GIS@Development*, vol. 8, issue 1, pp. 31:**

"One of the major decisions taken was to reorganize Survey of India to perform the role of a nodal agency in National Spatial data Infrastructure (NSDI). The present system of circles and zones was disbanded and each state now has a Spatial Data Centre created at its headquarters. Now, the Survey of India's presence will be in each district headquarters. In addition, there is going to be revolution in the way maps and digital map shall be handled in this country with the new map policy, which will be presented shortly. There will be series of maps, one exclusively for the Defence and other one for the civilians."

18. **Hadley, Clare & John Hammond (2003), Survey of India: A view from abroad, *GIS@Development*, vol. 7, issue 2, pp. 29-31:**

"The Survey of India and Ordnance Survey were formed in 1767 and 1791 respectively as military organisations for similar strategic reasons. Whilst the discipline of military doctrine was hugely successful in the earlier years with strict processes and consistent controlled procedures creating products renowned for their detail and accuracy, the legacy of that approach has left organistaions that are seen as inflexible, bureaucratic, and slow to respond, with a lack of clear focus on the evolving requirements of today's customers. One part of the SOI response to this challenge was a Workshop held during November 11-15, 2002 at Jim Corbett Park, Uttaranchal, India entitled 'Survey of India: Towards a Contemporary Renaissance'."

19. **Arunachalam, B. (2006), Indian prelude to British cadastral and revenue maps, *Coordinates*, vol. II, issue 3, pp. 22-5:**

"In the early decades of the 19th century, under the regulations of William Lambton and later George Everest, the primary and secondary triangulation networks began taking shape with minimal linear measures and precise angular measures to cover the whole country with a network of triangles, interconnected. The priority given by the Company Directors and the Indian Surveyor-Generals no doubt provided a pivotal role in unification and helped in creating an image of imperial space, unique and precise in the world. This system imposed from above did not contribute in any manner to the build up of a coordinates revenue and cadastral survey system in the country at the grass root level, and then building it up. The twain shall never meet, in the decades to follow proceeding on different planes and the Survey of India gradually lost sight of one of its primary tasks leaving it in hands of talatis, revenue inspectors and collectors. The cartographic anarchy was complete and Surveyor Generals combined into one group, who unfortunately never realized the importance of ground level surveys."

* * *

"Prime Minister Churchill, wise and clear through his words were, asked us to look at a map while he was speaking, for the map assisted him in telling what he could not put wholly or satisfactorily into words. Maps are one of the oldest and best forms of expression known to man."

In Kozakis, *World Topographical Map Coverage,* p. 553.

Appendices

APPENDIX I

Letter from Surveyor General's Office

Speed Post

भारतीय सर्वेक्षण विभाग
SURVEY OF INDIA

Save Paper.
Save Trees.
Save the World.

टेलिफैक्स
Telefax   +91-135-2744064, 2743331

वेबसाईट
Website   www.surveyofindia.gov.in

ई-मेल   sgo.soi@gov.in
E-Mail sgo.technical.soi@gov.in

भारत के महासर्वेक्षक का कार्यालय
Office of the Surveyor General of India
हाथीबड़कला एस्टेट, डाक बक्स सं. 37
Hathibarkala Estate, Post Box No. 37
देहरादून — 248 001 (उत्तराखण्ड), भारत
DEHRADUN - 248 001 (UTTARAKHAND), INDIA

No. T-1556 /1750-Policy   Dated: 31 May, 2019.

To

The Secretary to the Govt. of India,
Ministry of Science & Technology,
(Department of Science & Technology)
Technology Bhavan, New Mehrauli Road,
New Delhi – 110016.

[Kind Attn: Dr. Bhoop Singh, Head (NRDMS)]

Sub :-   **Writing Memoirs of Survey of India by Dr. Prithivish Nag, Former Surveyor
General of India – Reg.**

Ref: Letter No. NRDMS/11/001/2019, dt. 15.04.2019.

Sir,

    I have the honour to invite Department's kind attention to their letter cited under reference above. In the subject matter, it is submitted that Survey of India has witnessed and undergone major changes since independence and these changes, developments and transformation has not been documented systematically. Dr Prithivish Nag was Surveyor General of India and he has great insights about these significant historical events of Survey of India. Hence it is recommended to support this activity of recording these facts and information.

    This is for your kind information please.

    This letter has been issued with the approval of Surveyor General of India.

Yours faithfully,

(Pankaj Mishra)
Deputy Director,
for Surveyor General of India

D :TECH SECTION May, 2019\ Page135of 178-01-01-2008

APPENDIX II
Primary & Secondary Triangulation/Traverse Series

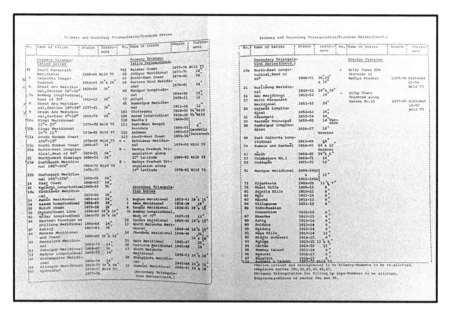

APPENDIX III
Conventional Charter of Duties

(a) Provision of geodetic control network, geodetic and geophysical surveys and accurate fixation of outlying islands through geodetic triangulation, trilateration and satellite geodesy.

(b) Provision of topographical cover in surveying and mapping for the entire country to meet the national requirements, including those of defence forces.

(c) Tidal predictions for 44 ports in the Indian Ocean, Arabian Sea and the Bay of Bengal including ports in Myanmar, Iran, Sri Lanka and Sultanate of Oman in the interest of good neighbourly relations.

(d) Compilation/mapping and production of geographical maps e.g., Railway Map, Road Map, Political Map, Physical Map etc.

(e) Preparation of the International Map of the World (IMW) series and the World Aeronautical Charts (WAC) series as a commitment to the International Civil Aviation Organization (ICAO).

(f) Surveys for development projects, e.g., power and irrigation, mineral exploration, urban and rural development etc.

(g) Surveying and mapping of forests areas, large cities and preparation of guide maps of cities/towns/places of interest.

(h) Surveying and mapping of Cantonments, surveying and mapping for aeronautical maps/charts for the IAF.

(i) Standardization of geographical names based on phonetics and participation in the international body set-up by the United Nations for this purpose.

(j) Demarcation of the external boundary of India, its correct depiction of maps published within the country. Also advising the Government of India on the demarcation of inter-state boundaries.

(k) Training officers and staff of the department, trainees from other central and state government departments and trainees from foreign countries.

(l) Promotion of Research & Development activities in the field of geodesy, photogrammetry, thematic mapping, production of three-dimensional maps, digital cartography and printing techniques etc.

(m) Introduction of modem technology in the related fields and indigenisation of equipment as an aid to import substitution. This includes development of instruments/materials indigenously to increase self-reliance and reduce the drain on the foreign exchange reserves.

(n) Coordination and control in providing aerial photographic cover for the whole country.

(o) Collaboration with training organizations, educational institutions and scientific bodies on specific projects to promote research and developmental activities.

(p) Representation at various international and national conferences to promote the growth of surveying and cartography and to introduce the state-of-the-art technology for optimum results.

(q) Support to Third World countries e.g., Nigeria, Afghanistan, Kenya, Iraq, Nepal, Sri Lanka, Zimbabwe, Indonesia, Bhutan, Myanmar and Mauritius etc. by providing technical know-how and expertise in various disciplines of surveying and survey education.

In addition, the Surveyor General of India acts as an adviser to various ministries of the Government of India on all surveying and cartographic matters. Survey of India also renders advice on the specifications of surveys and furnishes necessary data/maps to various central and state government departments for development, planning and defence purposes.

APPENDIX IV
Additional Charter

Due to advancement in information technology and flexible requirements of users of digital topographic data in a structured mode has forced the National Mapping Organizations (NMOs) world over to redefine their roles so as to fulfil the ever-increasing aspirations of data user community. The following items should be included in the charter of duties and responsibilities of SOI:

i) To provide leadership in matters related to NSDI and GSDI.

ii) As an NMO to fulfil his obligations to *International Steering Committee for Global Mapping*.

iii) Under the initiative of NSDI, undertaking necessary steps to develop *National Spatial Data Exchange Format* to include all the features of National Exchange Format for *Digital Vector Data* (DVD) launched by SOI in early nineties.

iv) Creation of boundary database up to the level of revenue village on 1:25,000 scale and its accessibility to public through NSDI clearing house.

v) Creation of *Digital Topographic Database* on 1:25,000, 1:50,000 and 1:250,000 scale and its accessibility to public through National Spatial Data Infrastructure or NSDI clearing house.

vi) Creation of Digital Cartographic Database of geographical maps to take part in the international initiative of *Digital Earth*.

vii) Creation of an organized digital database of geographical or Gazetteer names with a public access mandate.

viii) Creation of large-scale digital database coverage for metros and cities.

ix) Efforts must be made in the direction of integration and linkage of cadastral surveys with the *National Control Survey Framework*.

APPENDIX V

Properties of Survey of India

1.	Andhra Pradesh Hyderabad (IISM)	Hyderabad Kondapur 23 Km from Uppal	Plot one in Bhongir Camp, Nalgonda Distt. Kondapur Camp Range Reddy Distt.	1-Plot No.1 is 21772 Sq. mtr. (5.38 Acres) 2-Plot No.2 is 60702 Sq. mtr. (15.00 Acres) 3-4.02 Acres	196.19 Acre
2.	Assam & Nagaland GDC	Guwahati	Dakhingaon area Guwahati	7525.146 Sq.m (10 Bighas)	
2. A	Delhi (Western Printing Group)	New Delhi	Near 13 B.R.D Delhi Cantt. Svy. No. 120/1	1.628 Acre	5.872 Acre
3.	Gujarat, Damn & Diu GDE	Gandhi Nagar			20500 Sq. mtr
3. A	J&K GDC	Jammu	Nagrota		
4.	Jharkhand GDC, Ranchi	Ranchi	Chuka Toli, Doranda, Ranchi	0.76 Acres (Plot 1 is 0.68 Acres & Plot 2 is 0.08 Acres)	6.48 Acre
4.A	Karnataka GDC	Bangalore	Sarjapur road, Cormangalam		20 acre and 10 Guntas 9081.69 Sq. mt. Office Building 1198.71 Sq. mt. Residential building
5.	Kerala & Lakshadweep GDC	Thiruvananthapuram	Mannarkona, Vattiyoorkavu, Thrivandrum	8093.7 Sq. m. (2 Acre)	
6.	Madhya Pradesh GDC	Jabalpur	Plot No. 173 & 175 to 182 in, Scheme No.5, Vijay Nagar	6725 Sq. mtr. (No vacnt area as the construction of office building was tken up some columns were erected thereafter, work was	442477 Sq. ft.

7.	Maha-rashtra and Goa GDC	Pune	PhuleNagar Yeawda, Pune	No vacant land is available as it is within campus	41 Acre 17 Gunta
8.	Megha-laya & AP GDC	Shillong	Malki		13.519 Acre
9.	Odisha GDC Bhuba-neswar	Bhuba-neswar	Plot No. R1, Drawing No. 7684 in Mouza Jagamara, Khandagiri, Bhubaneswar	22795.9 Sq. mtr. (Ac 5.633De.)	6.847 Acre
10.	Panjab Haryana &Chandi-garh GDC & HP GDC	Chandi-garh	Sector-32, Chandigarh		12454.53 Sq. Yards 14380.91 Sq.Yards
11.	Rajasthan GDC (Ajmer Wing)	Ajmer	Arun Lal Sethi Nagar Yojana of UIT, Ajmer	12140.30 Sq. m.	
12.	Rajasthan GDC	Jaipur	Sec-6 Vidyadhar Nagar, Jaipur	7747.21 Sq.m.	
13.	Tamil-nadu, Pie & ANI GDC	Chennai	Village Velachery Taluka Mambal-amguindy	42.09.59 Sq. Mt.	
14.	Uttara-khand	Musso-orie	Landour Bazaar		773514 Sq. m.
15.	Uttar-khand	Dehradun	Hathi barkala Estate, GBO Compound and Survey Camping Ground, Rajpur Dehradun, Punch House	7530 Sq. M.	1445178.66 Sq. M
16.	UPGDC, Lucknow	Lucknow	Gomti Nagar Scheme Phase-1 & Phase 2	Plot No.2 is 21,777 Sq. mtr.	Plot No. 1 is 2,492.5 Sq. mtr. 1331.88 Sq. Mt. (Picup)

APPENDIX VI
Posting of Officers after Reorganisation in 2003

	State directorate	Headquarters	Director
1	Andhra Pradesh	Hyderabad	Brig Manoj Tayal
2	Aurnachal Pradesh-Meghalaya	Shillong	Brig B.D. Sharma
3	Assam-Nagaland	Guwahati	Shri B. Mahapatra
4	Bihar	Patna	Shri Somara Tirkey
5	Jharkhand	Ranchi	Brig R.N.B. Verma
6	Orissa	Bhubaneshwar	Shri C.R. Biswas Shri U.N. Gurjar
7	Chhattisgarh	Raipur	Shri Hari Om Prasad
8	Karnataka	Bangalore	Brig N.R. Ananth
9	Goa	Pune	Brig P.K. Chaudhary
10	Gujarat-Daman & Diu	Ahmedabad/ Gandhinagar	Shri Ashok Prim
11	Haryana	Chandigarh	Shri S.M. Pal
12	Punjab – Chandigarh	Chandigarh	Maj Gen Wirk
13	Himanchal Pradesh	Chandigarh	Brig C.S. Bewli
14	Jammu & Kashmir	Jammu	Brig Bahl
15	Tamil Nadu	Chennai	Shri M. Dharam Raj
16	Kerala-Lakshadweep	Bangalore	Brig M.V. Bhat
17	Madhya Pradesh	Jabalpur	Maj Gen Dr B.C. Roy Dr R.S. Madame
18	Maharashtra	Pune	Maj Gen V.N. Nerikar Lt Col P.S. Samudra
19	Rajasthan	Jaipur	Brig Alok Kumar
20	Uttar Pradesh	Lucknow	Shri S. Subba Rao
21	Uttaranchal	Dehra Dun	Brig S.K. Pathak
22	West Bengal-Sikkim-Andaman & Nicobar Is.	Kolkata	Shri T.K. Bandho-padhya
23	Kolkata Printing Group	Kolkata	Shri C. Chakravarty

	State directorate	Headquarters	Director
24	Hyderabad Printing Group	Hyderabad	Shri P. Nagendra Prasad
25	Dehli Printing Group	Delhi	Shri R.M. Tripathi
26	Dehra Dun Printing Group	Dehra Dun (MPO)	Shri G.C. Bairagi
27	Modern Computer Centre	Dehra Dun	Brig Girish Kumar
28	Digital Mapping Centre	Dehra Dun	Brig P.K. Vachher
29	Digital Mapping Centre	Hyderabad	Brig T.P. Malhotra
30	Business & Publicity	Dehra Dun	Shri Madhwal
31	Director Survey (Air)	Delhi	Brig G.S. Chandela
32	Geodetic & Research Branch	Dehra Dun	Brig Dr B. Nagrajan
33	Research & Development	Hyderabad	Brig V. Singhal
34	National Spatial Data Infra-structure	Delhi	Brig Dr R. Siva Kumar
35	Survey Training Institute	Hyderabad	Maj Gen M.G. Gopal Rao
			Brig V.R. Mahendra
			Brig KRMK Babaji Rao
			Shri S.J. Vaidhya
			Shri K.R. Meena
			Shri BP Nainwal
36	SGO (Delhi)/Boundary Cell	Delhi	Dr M.C. Tiwari
37	SGO (Dehra Dun)	Dehra Dun	Maj Gen V.C. Tyagi
			Shri S.P. Geol
			Brig Girish Kumar
			Shri Ram Prakash
			Shri Chandra Pal
			Shri Naveen Tomar
38	Delhi	Delhi	Shri C.B. Singh
39	Mizoram-Tripura-Manipur	Silchar	---

APPENDIX VII
Welcome Address by Dr Prithvish Nag

Welcome address by Dr Prithvish Nag at the opening of the "The Great Arc Exhibition" Atlantis Gallery, Trueman Brewery, Brick Lane, London. 7th October 2003.

"I am extremely delighted to address this august gathering which has assembled in order to take part in the celebration of the Great Arc function in London jointly organized by the Department of Science & Technology presence of our Hon'ble Minister of. Science & Technology, Government of India, Representatives from the Indian Mission and the Nehru Centre, both in London. It is an important occasion for the Survey of India as well. On its 236th year of its existence, this institution has made its presence not only in India but also all over the world. But the more important feature is its scientific contribution, which was somehow not given due importance. One such initiative was to recognize the importance of the *Great Trignometrical Survey* or *Great Arc* in short. About two hundred years ago our attempt to accurate measure the country led to the formation a strong backbone in mapping the country which ultimately paced way for the establishment of several specialized survey institutions in the country earlier, and development of infrastructure recently. Nevertheless, the Great Arc celebration was an occasion for introspection and for future initiatives. In last one and half year or so we took up several activities, some of which I would like to mention here:

Survey of India is the premier national survey and mapping organization in the country. In its assigned role as a national mapping agency, it bears a special responsibility to ensure that the country's domain is explored and mapped suitably to provide base maps for expeditious and integrated development of the nation. In the process, Survey of India has been producing all-purpose topographical and various other series maps required by defence, general administration, internal security, developmental needs, irrigation, watershed management, resource management and various types of engineering projects. It is also responsible for establishing precise planimetric control, heights above MSL, gravity, geomagnetic and tidal prediction as prerequisite for mapping activities and other scientific applications. The department is committed to provide technical expertise to other countries in the field of geodesy, surveying, cartography and survey education.

With the introduction of digital technology in the department, digital topo-graphical database for entire country is being created in various planning also developing processes and creating GIS. Its specialized directorates such as Geodetic & Research Branch, Research and Development Directorate and Survey Training Institute have been further strengthened to meet the growing requirements of user community. The department has also been immensely contributing to a number of multi-institutional scientific programmes related to the field of geophysics, remote sensing, glaciology, study of seismicity and seismotectonic, scientific expedition to Antarctica and digital data transfer.

On this occasion we had a series of meetings, workshops and discussions for re-engineering this old but vital institution. Outside experts, even from abroad, were invited. These gatherings helped us in taking some new initia-tives as a part of the celebrations. Apart from bringing out publications - one of which is to be released here, conducting treasure hunts in different parts of the country, conducting neighbourhood mapping programme, highlighting our activities in the media, seminars, exhibitions and the like, serious scientific programmes were also taken up.

Survey of India has taken up major programmes related to technological devel-opment in the field of mapping activities with the initiatives of the Department of Science & Technology. The National Spatial Data Infrastructure (NSDI) and new series of maps have been initiated under this programme.

India is fast moving into information and knowledge society. Emphasis is increasingly placed on IT driven "transparent" e-governance. The nation has been generating voluminous field/ digital data *i.e.,* information through systematic topographical surveys, geological surveys, soil surveys, cadas-tral survey etc. Access and availability of such information to the citizen, society. private enterprises and government are of immense importance. As a part of this vision, NSDI is being involved through partnership approach among various data generating agencies to facilitate integration, easy access and networking of databases. NSDI has been conceived as a national system that synergistically combines the resources and infrastructure of various players with the power of IT and enabling information support for decision making in government, industry, academia and other organisa-tions besides serving the public needs.

NSDI strategy and action plans have been formulated through multi-institutional approach involving Department of Science & Technology (DST), Survey of India (SOI), Indian Space Research Organisation (ISRO), Geological Survey of India (GSI) and National Atlas & Thematic Mapping Organisation (NATMO) etc.

In order that the maps in analogues and digital form are available to all users in India for sustainable development of the country, a new series of maps on WGS 84 datum is planned and the work on various components involved in this proposal has already commenced.

Great Trigonometrical Survey of India in over two and half centuries collected valuable information for the service to the humanity and science. Its products and wealth of information are being utilized in the development of the nation. Geodetic and Geophysical information like planimetric and height control, gravity, geomagnetic and sea level data collected with the aim of serving humanity by forewarning them for the natural disasters and by bringing the information of resource availability to the earth scientists.

Today's programme in London is in continuation with the Great Arc celebrations of Survey of India. We have utilized this occasion to demonstrate our achievement and expertise in the field of surveying, mapping and related fields. Perhaps when we meet again on similar events, we will have much more to talk about us."

Inaugural Address by Dr. Murli Manohar Joshi, Minister for Human Resources Development, Science & Technology and Ocean Development, Government of India at the opening of the "The Great Arc Exhibition" Atlantis Gallery, Trueman Brewery, Brick Lane, London. 7th October 2003.

"On occasions such as this, it is customary to begin speeches by recalling historic relations between countries and civilisations. In most places this is nothing more than mouthing the clichés of diplomatic parlance. Between India and UK. however, the nature of the relationship between our two civilisations is so redolent with history, so pregnant with 'memory and desire' (to use an Eliotic turn of phrase) that even clichéd expressions acquire profundity and intensity of meaning. I doubt if there are any other two civilisations, so different and yet so closely intertwined, in terms of people- to-people contact, language, cuisine and above all the mind and the intellect. Today, we celebrate an aspect of this relationship.

In hindsight, the Great Arc odyssey is capable of being viewed in many different ways and at many levels. Conventionally, it has been seen as the mapping of an empire a typically grandiose, colonial enterprise to consolidate the military, economic and territorial gains of an imperial state. Scratching beneath the surface, we begin to see the magnificent obsession of a few individuals, using the cloak of imperialistic ends to justify their fascination with a particular branch of science and an almost fetishistic passion for measurement and empirical observation. If we go a little deeper we can see the unfolding of a still larger, grander. story that of a unique partnership of minds. We need to see through the colonial context and the asymmetry of political power at that time, to appreciate the collaborative nature of the project and the strength of the team effort which went into it. The Survey was not a military campaign with star British generals leading an army of passive Indian subalterns and foot soldiers. Nor was it an industrial enterprise which could employ armies of cheap Indian wage labourers regimented into an assembly-line order. The utilitarian ends of the exercise were not apparent enough. There was no pot of gold or buried treasure at the end of it. Yet, thousands and thousands of men offered their skills, their brainpower, their brawn, their patience and indeed their lives, for a scientific pursuit. For half a century this gigantic collaborative scientific mission required people to battle the most hostile of terrains, the fiercest of tropical jungles, the worst fevers and epidemics, as well as intellectual challenges and complexities which were truly mind- bending. All this for a map, for a measurement, for knowing just a little better the figure of the earth! In the history of science there is no parallel to this endeavour in terms of its scale and magnificence or in terms of a giant partnership in the sheer joy of science. A partnership where the British brought their societal attributes of a passion for empirical detail, for precision. for intellectual discipline and physical challenge and the Indians their 'genetic software' for mathematical, astronomical and spiritual insights, for logical disputation and for philosophical abstraction. The convergence is unique, the synergy achieved possibly unmatched. If we have to look for an inspirational model of international cooperation in a scientific endeavour, a model of Werner Heisenberg's desire to provide an Eastern spiritual compass to the comforts of the Western technological ship, a model of multiple disciplines and multiple skills converging for a scientific end, it will be difficult to find another so metaphorically rich.

In a world which refuses to see beyond narrow economistic, technocratic and consumerist ends. it is difficult to believe how a scientific exercise of this magnitude could inspire such multitudes. I believe that this could only have happened in India. In the long history of civilisations, the Indian civilisation is one of the few for whom the scientific impulse to know, to enquire has been a defining feature of its existence. In most other societies the realm of scientific inquiry was appropriated by a privileged few. Others were expected to follow the governing metaphor being the shepherd and his flock. In our case, we adopted the restless pursuit of creative enquiry as our religion. The task, the *dharma* of each individual was to pursue, through a breathtaking choice of methodological options, the ends of true knowledge and enlightenment.

In the exhibition that you will see today we have gone beyond the story of the Great Arc to give you a glimpse into the extraordinary nature of the Indian scientific tradition. We have tried to show two aspects of this tradition. One, that the spirit of conscious scientific enquiry is a part of every segment of Indian society and is not limited to those who are fortunate enough to attend a university course in it. That is why we could have a

lowly clerk like Ramanujan dedicate his total life to the pursuit of mathematical abstractions. A Radhanath Sikdar who became George Everest's,
'Chief Computer' and without whom the Great Arc could literally not
have achieved the heights of Mount Everest. A Syed Mir Mohsin Hussain
- a watchmaker who went on to engineer, on his own. the most complex
mathematical instruments, some of which are on display here. The pundits
- Nain Singh and Kishen Singh ordinary schoolteachers. in a vernacular
tradition who became pioneering surveyors and geographers. Two, the
Indian scientific tradition is non-dichotomous and does not place science
in a compartment separated from society, from philosophy, religion,
aesthetics, and ethics, from the arts and the humanities. We have. therefore,
chosen what the West often, through its binary mindsets, views as the polar
opposite of science - the arts - to show that Indian art traditions do not see
themselves as alienated from scientific concerns. Through their own idiom,
their own grammar, they reflect on the basic issues of science I creation
and its origins, astronomical phenomena, the concept of 'zero' or infinity,
natural cycles, environment, the magic of numbers, of permutations and
combinations, and distance, directions, and measurement. It is the depth,
penetration, and the richness of the Indian scientific tradition that we have
tried to give you a glimpse of.

Let me give you a brief example from our ancient texts of the importance
attached in tradition to scientific knowledge. In the *Mundaka Upanishad*, an
earnest student asks this question of a great teacher:

> *Kasmin no bhagavo vijnate sarvam idam vijnatam bhavati?*

What is that reality. O Blessed one, by knowing which we can know all that
there is in the universe?

To this, the teacher gives a surprising reply:

> *Dve vidye veditanye iti ha sma yat Brahmavido radanti para chiva apara ca.*

There are two types of knowledge to be acquired by men, so say the knowers
of Brahman. One is called *parauidya* or higher knowledge; the other is called
aparavidya or lower knowledge.

The Upanishad goes on to say that both types of knowledge are critical.

Lower knowledge consists of the sacred Vedas themselves, phonetics, the code of rituals, grammar, etymology, mathematics, prosody and astronomy. The category of *paravidya* is described thus:

Atha para yaya tat aksaram adhigamyate.

That is para by which the Imperishable is realized.

The burden of the argument is that, for reaching a state of total realisation, the path of lower knowledge, e.g., scientific knowledge has to be traversed before one attempts to go beyond to higher knowledge. In other words, Indian religious thought does not require a separation or dichotomy between scientific reasoning and religious belief but, in fact, treats it as an essential route to higher forms of knowledge. The scientific spirit of questioning, of searching is integrally built in into the process of religious discovery. Science is, therefore, to be seen in the true spirit of its root Latin meaning - to know.

Ladies and Gentlemen, we have been asked why we are celebrating the Great Are Survey with so much fanfare. We believe it is a metaphor of India's leadership in science in the past, in the present and in the future. Many in the West associate India with primitivism and backwardness. We wish to show you that, on the contrary, it is the extraordinary sophistication and richness of Indian science which accords us the status of a knowledge superpower. Today, at the back of every significant development in infor-mation sciences, bio-sciences. life-sciences, nano-sciences, anywhere in the world, there is an Indian scientist lurking. The field of geo-spatial sciences - the focus of this exhibition - is one of the many where we have a lot to offer. This exhibition is also a metaphor for science to inspire international partnerships to move to a world that is driven by the ideals of sustainability in consumption and production, and in which the joy of scientific enquiry infects every human being on this earth. I hope the experience that we have tried to create will have done some justice to what we intended to do. I wish you all an exciting viewing."

APPENDIX IX
Survey of India a Glance, 2002-05

EXISTING ROLE OF SURVEY OF INDIA

1. Topographical surveying and mapping

2. Geodetic control

3. Tidal data collection

4. Gravity and geomagnetic observations

5. Aeronautical charts and tables

6. International boundary

7. Research development and training

8. Project surveys and special surveys

9. Geographical maps

PRESENT DEMAND

1. Digital topographical data

2. Current Topographical Maps

3. NSDI activities

4. Infrastructure maps including of urban areas

REQUIREMENTS FROM INDUSTRY

1. To become front agency for bargaining projects: national and international

2. To provide fundamental data sets in digital form for value addition

3. To jointly bid for projects etc.

4. To bundle SOI data with software

5. To certify data quality

STATUTORY PROVISIONS

1. Staff reduction

2. Higher financial provision for modernization

3. Commercialization

4. Outsourcing

5. Revenue generation

FANANCIAL SCENARIO

1. About 90 per cent of the annual expenditure goes on non-plan sector out of which maximum is on "Salary" head

2. Revenue generated every year is about 20 per cent

3. Uneconomic small units all over – consolidation required

4. Reasonable sum is paid for rents and taxes for hired buildings

5. Large estates are underutilized

6. Surplus staff is some sectors

EMERGING SCENARIO

1. Introduction of new technologies

2. New Products

3. New partnerships

4. New vision

VISION

The SOI will take a leadership role in providing customer focused, cost effective and timely geospatial data, information and intelligence for meeting the needs of sustainable national development and new information markets.

OBJECTIVES

To provide and maintain the NDTDB conforming to national standards including national foundation data set:

1. The national spatial reference frame

2. The national DEM

3. The national topographic template

4. Administrative boundaries

5. Toponymy

MISSION

SOI dedicates itself to the advancement of theory, practice, collection and applications of geospatial data, and promotes an active exchange of information, ideas, and technological innovations amongst the data producers and users who will get access to such data of highest possible resolution at an affordable cost in the near real-time environment.

AIMS

1. SOI will be vibrant, *i.e.,* respective to the needs of user community in India.

2. To promote professionalism, incorporating high standards of ethics and conduct among geospatial data producers.

3. To provide a forum for all the professionals to share knowledge and expertise.

4. To stimulate, encourage and participate in research, development and applications of geospatial data.

5. To be active in all the areas of applications of geospatial data through special interest groups which will be dynamic to accommodate the emerging technologies and applications.

6. To contribute to the formulation and organizing educational and training programmes needed by professionals, users and the public.

7. To actively participate and advise the public bodies including the Government in the formulation of policies affecting the professions of surveying and mapping.

GOALS- LONG TERM

To convert SOI into completely digital environment by 2005.

TARGETS – SHORT TERM

1. All maps on scales 1:50,000 and 1:250,000 will be digitized by 31 December 2003

2. NSDI should be on the net giving the availability of data and metadata by 15 December 2002.

3. Establishment of GDCs in all States by 31 December 2003

4. Introduction of new products and services by 01 Dec 2002 including dual series maps

5. Recruitment of scientists from various disciplines by 01 July 2003

6. Induct new technologies ALTM and establish supremacy in geospatial data acquisition, management and dissemination.

7. To introduce new technologies in field work by 01 Jul 2004

8. To take full advantage of the new and evolving ICT methods for storage, analysis, conversion and dissemination.

IMMEDIATE REQUIREMENTS

1. More powers to Surveyor General regarding:

2. Disciplinary action

3. Transfer, promotion, placement

4. Collaboration with industry

5. Publicity

6. Out-sourcing

7. Engagement of consultants on higher salaries

Appendix X
Court Cases, 1998-2003

Year	Case B/F	New cases	Cases decided	Pending
1998	164	41	38	167
1999	167	34	44	157
2000	157	26	17	166
2001	166	57	59	164
2002	164	60	40	184
2003	184+4*	69	52	201+4*=205

*Contempt cases

Acknowledgements

- Shri Ram Naik, Former Governor of Uttar Pradesh.

- Dr Murli Monohar Joshi, Former Union Minister for Science & Technology and Ocean Development; and Human Resources Development.

- Shri Kapil Sibal, Former Union Minister for Science & Technology and Ocean Development; and Human Resources Development.

- Late Shri Bachi Singh Rawat, Former Union Minister of State for Science & Technology.

- Professor V.S. Ramamurthy, Former Secretary, DST

- Dr K. Kasturirangan, Former Chairman ISRO & Secretary, ISRO.

- Dr K. Radhakrishnan, Former Chairman ISRO & Secretary, ISRO.

- Dr Shailesh Nayak, Former Secretary, Ministry of Earth Sciences.

- Professor T. Ramasami, Former Secretary, DST

- Professor Ashutosh Sharma, Former Secretary, DST

- Professor Kalpalata Pandey, Former Vice-Chancellor, JNC University, Ballia

- Shri Amitabha Pande, Former Joint Secretary, DST

- Late Shri Sanujit Ghose, Former Joint Secretary, DST

- Lt. General Girish Kumar, Former Surveyor General of India

- Dr Swarna Subba Rao, Former Surveyor General of India

- Shri Naveen Tomar, Former Surveyor General of India

- Shri Sunil Kumar, Joint Secretary, DST & Surveyor General of India

- *Following officers of the Department of Science & Technology, Government of India:*

  Dr Akhilesh Gupta, Senior Advisor; Major General (Dr) Siva Kumar, Former Advisor/Head, NRDMS & CEO, NSDI; Dr Bhoop Singh, Former

Head, NRDMS & NSDI; P.S. Acharya, Former Head, NRDMS & NSDI; Y.S. Rajan, Former Executive Director, TIFAC & Advisor; Dr Debapriya Dutta, Head, SEED; Advisor, National Geospatial Programme; Dr A.K. Singh, Former Scientist, National Geospatial Programme; Dr Shubha Pandey, Scientist, National Geospatial Programme; A. J. Kurian, Former Director, SMP Division; Shambhu Singh, Former Director; Ashok Harnal, Former Director; Shri Avinash Dikshit, Former Director; Kamal Prakash, Former Under Secretary; S.P. Katnauria, Former Under Secretary; A.C. Mullick, Former Under Secretary; B.K. Raichandani, Former Under Secretary; Anuj Kumar Tripathi, Principal Staff Officer, DST.

– *Following officers of Survey of India:*

Maj Gen Manoj Tayal, Maj Gen B.D. Sharma, B. Mahapatra, Col R.N.B. Verma, Maj Gen N.R. Ananth, Maj Gen P.K. Chaudhary, Ashok Prim, S.M. Pal, Maj Gen Rajesh Bahl, Maj Gen C.S. Bewli, M. Dharam Raj, Maj Gen M.V. Bhat, Late Maj Gen Dr B.C. Roy, Late Dr R.S. Madame, Brig P.S. Samudra, Maj Gen Alok Kumar, Maj Gen S.K. Pathak, T.K. Bandhopadhya, C. Chakravarty, P. Nagendra Prasad, Late R.M. Tripathi, G.C. Bairagi, Maj Gen P.K. Vachher, Maj Gen T.P. Malhotra, H.B. Madhwal, Maj Gen G.S. Chandela, Maj Gen (Dr) B. Nagrajan, Maj Gen V. Singhal, Maj Gen M.G. Gopal Rao, Maj Gen V.R. Mahendra, Maj Gen KRMK Babaji Rao, S.J. Vaidhya, K.R. Meena, B.P. Nainwal, Maj Gen V.C. Tyagi, S.P. Geol, Ram Prakash, Late Chandra Pal, C.B. Singh, C.R. Biswas, U.N. Gurjar, Somara Tirkey, Rajive Srivastava, Maj Gen V.P. Srivastava, Dr S.K. Singh, Nitin Joshi, C.K. Mamik, Ram Prasadd, R.K. Nim, P.K. Chaudhury, V.K. Gupta, Maj Gen A.K. Sinha, Maj Gen A.K. Padha, P.V. Rajsekhar, U.N. Gujar, Sanjay Kumar, T. Sanjeev Kumar, Maj Gen S Ravi, J.K.Rath, R.K. Nigam, S.K. Sinha, Dr M. Stalin, Maj Gen Umesh Chandra, Maj Gen Alok Kumar, Maj Gen Kali Prasad, O.P. Tripathi, Maj Gen R.C. Padhi, Maj Gen Anil Kumar, Nirmalendu Kumar, V. Ravichandran, M.I.Malik, Brig A.K. Rew, Maj Gen K.K. Naithani, Shri M. Dharamraj, Shri P.K. Chaterjee, Late D.P. Issar, Maj Gen Amrik Singh, Maj Gen P. N. Koul, Maj Gen V.N. Nerikar, Jaswant Rai, Late Maj Gen R.S. Tanwar, Maj Gen Shamsher Singh, Vipin Chandra, S.V. Singh, Maj Gen V.S. Deo, Brig K.G. Behl, Charanjit Kaur, Ram Prasad, Amar Dev Bahuguna, Rajesh Kumar, P.S. Chopra, S.K. Sinha, Pankaj Mishra, Dr K.K. Shukla, Col. Rajat Sharma, Swarnima Bajpai, Devbrata Palit, and Ashish Sanyal.

– *Following officers of NATMO:*

G.N. Saha, A.K. Malik, Late Dr A.K. Dasgupta, Dr Rajendra Prasad, Dr R.K. Dhabai, Dr B.P. Singh, Late Dr G.C. Debnath, Dr Baisakhi Sarkar, Chandra Singh.

– *Colleagues from the Indian Institute of Technology, Banaras Hindu University:*

Prof P.K. Singh, Dr Anuragh Ohri, Dr Sishir Gaur.

– *Other colleagues:*

Sanjay Kumar, Dr Ravi Gupta, B.M.L. Sharma, Rajesh Mathur, Vijay Kumar Singh.

– **Drivers:**

Manmohan Singh, Umesh Joshi, Mangal Singh, Kissan Lal, Daya Ram, Raja Ram and Chand Ratan Goswami.

Milton Keynes UK
Ingram Content Group UK Ltd.
UKHW041001231023
431168UK00002B/28

9 781804 412381